Mechanical Science for Technicians 3

Mechanical Science for Technicians 3

W. Bolton
Adviser, Technician Education Council

THE BUTTERWORTH GROUP

UNITED KINGDOM Butterworth & Co (Publishers) Ltd
London: 88 Kingsway, WC2B 6AB

AUSTRALIA Butterworths Pty Ltd
Sydney: 586 Pacific Highway, Chatswood, NSW 2067
Also at Melbourne, Brisbane, Adelaide and Perth

CANADA Butterworth & Co (Canada) Ltd
Toronto: 2265 Midland Avenue, Scarborough, Ontario M1P 4S1

NEW ZEALAND Butterworths of New Zealand Ltd
Wellington: T & W Young Building, 77–85 Customhouse Quay, 1,
CPO Box 472

SOUTH AFRICA Butterworth & Co (South Africa) (Pty) Ltd
Durban: 152–154 Gale Street

USA Butterworth (Publishers) Inc
Boston: 10 Tower Office Park, Woburn, Mass. 01801

First published 1980

British Library Cataloguing in Publication Data

Bolton, William, *b. 1933*
Mechanical science.
1. Mechanics, Applied
I. Title
620.1 TA350 79–41525

ISBN 0-408-00486-X

Typeset by Scribe Design, Gillingham, Kent

Printed in England by Page Bros Ltd, Norwich, Norfolk

Preface

This book is designed to meet the needs of technicians studying the *Mechanical Science 3* unit in the Engineering awards of the Technician Education Council (standard unit reference number U75/058). The book covers this unit and additionally includes revision material. The subject matter is developed through step-by-step arguments and discussions, liberally illustrated. Worked examples and problems are included.

The aim of the *Mechanical Science 3* unit is to develop the students' analytical techniques in the application of scientific principles to mechanical engineering situations, the unit being designed to be studied concurrently, or after, a study of basic engineering science. The book should thus be suitable for use in a wide variety of other technical or technician courses in many other countries where a similar course aim occurs.

The general objectives (i.e. teaching goals) of this book are that the student (the numbers in brackets refer to the comparable objectives in the TEC standard unit):

1. Force
Determines when forces are in equilibrium.
Determines the resultant of a number of forces.
Resolves forces into their components.
Determines the moment of a force.
Solves problems in static equilibrium.
Determines centres of mass and centroids.

2. Stress and strain
Uses the concepts of stress, strain, modulus of elasticity, yield stress, strength and Poisson's ratio in simple stress–strain problems involving both tension and compression of homogeneous materials (A1).
Calculates stresses and strains due to temperature changes for homogeneous materials (A1).
Solves, from first principles, problems involving composite bars under uniaxial loads (A1).
Solves problems involving shear stress, shear strain, modulus of rigidity and shear strength (A1).

3. Bending beams
Develops and uses the simple theory of the bending of beams (A2).

4. Torsion
Develops and uses the simple theory of the torsion of circular section bars, both solid and hollow (A3).
Solves problems involving the transmission of power by shafts.

5. Force and motion
Solves problems involving Newton's laws.
Solves problems involving the conservation of momentum.
Solves relative velocity problems.
Solves problems involving work, potential and kinetic energies (B6).

6. Angular motion
Solves problems involving angular velocity and angular acceleration (B4).

Solves problems involving torque and angular acceleration (B4).
Solves problems involving angular kinetic energy (B6).
Derives the expression for centripetal acceleration and solves problems involving centripetal acceleration and force, including banking (B4).

7. Simple harmonic motion

Describes the characteristics of simple harmonic motion oscillations and solves problems involving such motion (B5).
Explains what is meant by resonance (B5).

8. Fluids in motion

Describes the laws of pressure in static fluids.
Derives the equation of continuity and Bernoulli's equation, solving problems involving the equations (C7).
Calculates the force exerted on plates by fluid jets (C7).

Contents

1 Force

FORCE AND ITS UNITS

A *force* is commonly called a push or a pull. If the push or pull is applied to an object at rest it causes the object to start moving and accelerate. If the push or pull is applied to an object already moving the result is a change in velocity and thus an acceleration. The size of the acceleration of a body can be used as a measure of the force. The basic unit of force, the *newton* (N), is defined in terms of the acceleration produced for an object of standard mass. The newton is the force which produces an acceleration of 1 m/s^2 when acting on an object of mass 1 kg. Common multiples of the newton force are:

1 kilonewton (kN) = 10^3 N
1 meganewton (MN) = 10^6 N
1 giganewton (GN) = 10^9 N

Forces are sometimes described in terms of the gravitational force acting on a mass. An object of mass of 1 kg experiences a gravitational force of about 9.81 N, the precise value depending on the place on the earth where the mass is located. This force of 9.81 N is thus sometimes written as 1 kgf. Thus, a force of 5 kgf is the same size force as the gravitational force that acts on a 5 kg mass.

When a force acts on an object and causes it to accelerate, the direction of the acceleration depends on the direction of the force. Thus, for the effect of such a force to be forecast the size and the direction have both to be specified. Quantities for which both size and direction have to be specified for their effects to be forecast are called *vector quantities*. Those quantities for which only the size needs to be specified are called *scalar quantities*. Force is an example of a vector quantity, volume is an example of a scalar quantity.

Although the unit of force is defined in terms of its dynamic effect, i.e. the effect being in terms of motion, a force may be measured in terms of the deformation which it can produce in some object. Thus, if a force is applied to each end of a spring then the spring will be extended, or compressed. The amount of extension or compression can be taken as a measure of the forces involved.

FORCES IN EQUILIBRIUM

If an object is acted on by just one force then it will accelerate in the direction of the force. If, however, two forces act on the object it is possible that they might be of such a size and in such a direction that they result in no acceleration of the object. The forces in such a case are said to be in *equilibrium*. For two forces to be in equilibrium then they have to be equal in size and acting in exactly opposite directions (Figure 1.1). In addition the lines of action of the two forces must pass through one point, and when this happens the forces are said to be *concurrent*.

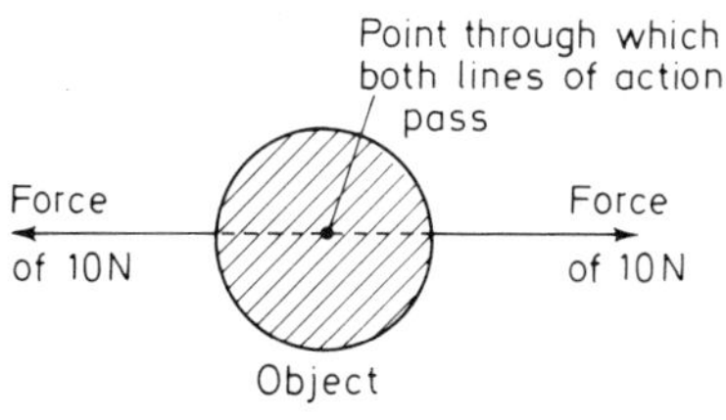

Figure 1.1 Two forces in equilibrium

A single force may be balanced, i.e. in equilibrium, with any number of forces. For three forces to be in equilibrium a number of conditions have to be satisfied:

1. they must all be in the same plane, i.e., *coplanar*,
2. they must have their lines of action all passing through one point, i.e. *concurrent*,
3. if the forces are represented in magnitude and direction by arrowed straight lines then these lines when taken in the order of the forces must form a triangle, known as the *triangle of forces.*

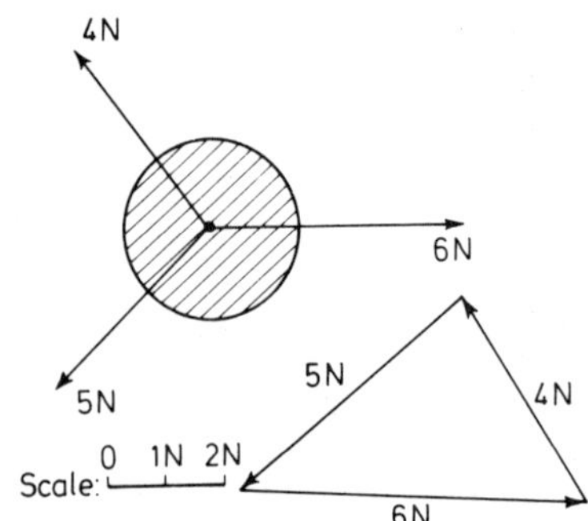

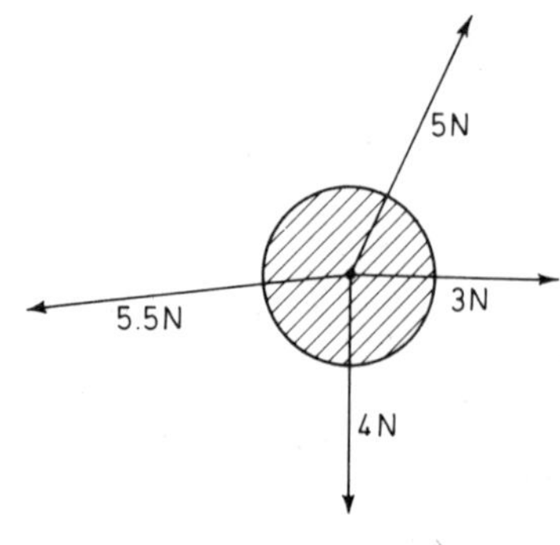

Figure 1.2 Three forces in equilibrium

Figure 1.2 shows an example of three forces, coplanar and concurrent, that are in equilibrium and the resulting triangle of forces. If the forces had not been in equilibrium then the forces represented by the lengths and directions of the sides of the triangle would not have exactly formed the closed triangle shape.

If more than three forces act at a point the triangle of forces is extended to become a *polygon of forces*. If the forces are concurrent and coplanar and if each force is represented in magnitude and direction by lines, then these lines when taken in the order of the forces must form a polygon, i.e. a closed shape. If the system of forces is not in equilibrium then the polygon does not close.

Figure 1.3 shows an example of four forces acting at a point and all in the same plane. Because the forces are in equilibrium the shape produced by representing each of the forces in size and direction by lines is a closed shape.

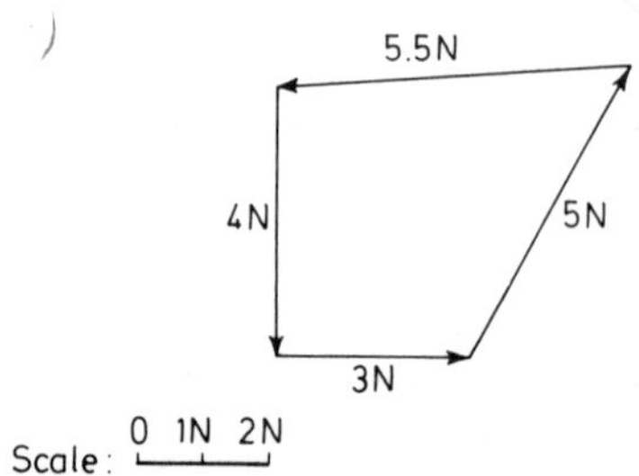

Figure 1.3 Four forces in equilibrium

Example 1 An object is suspended from a spring, as in Figure 1.4. The gravitational force acting on the object is 5 N. When the object is first attached to the spring, the spring extends. When the motion ceases the object is in equilibrium. What is the force exerted on the object by the spring when it is in equilibrium?

The gravitational force and the force exerted by the spring act in the same straight line and thus the force exerted by the spring must be the same as the gravitational force, i.e. 5 N.

Example 2 Figure 1.5(a) shows an object supported by two wires. If the object is in equilibrium what are the forces in the two wires?

With a scale of 10 N to 1 cm the 25 N force is represented by a line of length 2.5 cm. Drawing lines at the 60° and 30° angles to give the directions of the tensile forces and making them of such a length as to complete the triangle results in lines of length 1.2 cm and 2.2 cm (Figure 1.5(b)). This means that the forces are $T_1 = 12$ N and $T_2 = 22$ N to the accuracy with which the sides can be measured.

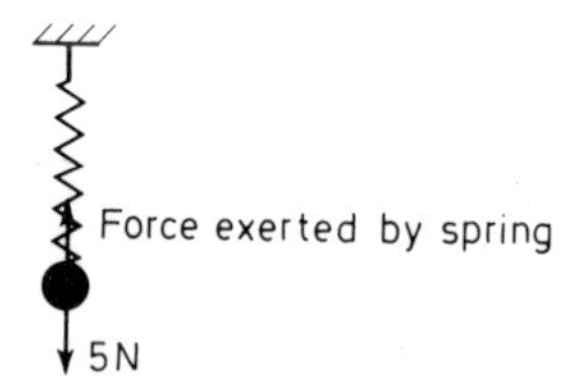

Figure 1.4

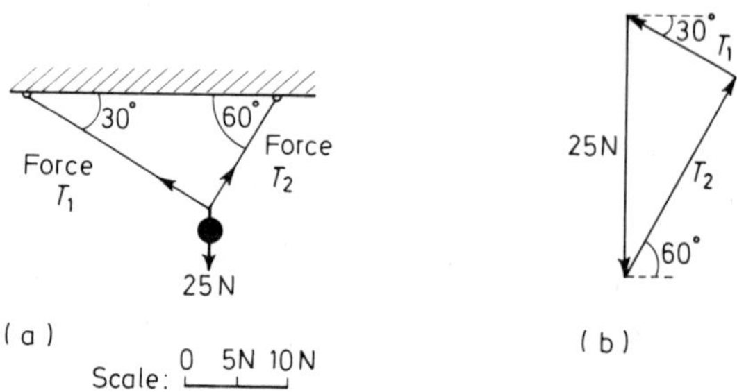

Figure 1.5

RESULTANT OF A SYSTEM OF FORCES

When an object is acted on by a number of concurrent forces, what is the effect of these forces? One way of deciding is to consider what single force could be used to replace the system of forces and then consider just that one force. A single force which is used to replace a system of forces and have the same effect as that system is called a *resultant force.*

The resultant force can be found by the use of the *parallelogram rule.* This rule can be stated as: the resultant force of two forces F_1 and F_2 is the diagonal of the parallelogram in which F_1 and F_2 are, when drawn to scale and in the appropriate directions, the adjacent sides.

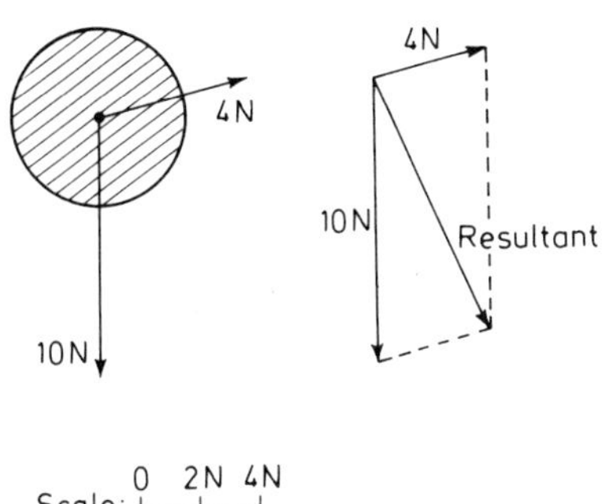

Figure 1.6 Finding a resultant force using the parallelogram rule

Figure 1.6 shows the parallelogram rule being used to find the resultant of two forces, the forces being 10 N and 4 N in the directions indicated. Lines are drawn to represent the two forces, the lines being in the same direction as the forces and drawn to scale. The dotted lines then complete the parallelogram. The diagonal, starting from between the two forces, is the resultant. In this case the resultant has a length of 2.4 cm and thus represent a force of size 9.6 N. The direction of the resultant is the direction given by the diagonal.

The parallelogram rule yields the resultant force acting on an object. If the object had been in equilibrium then the force needed to produce that equilibrium would be opposite and equal in size to the resultant. Thus instead of using the parallelogram rule to find the resultant we can find the force that would be needed to give equilibrium and then take the resultant to be in the opposite direction and the same size. Figure 1.7 shows both the parallelogram and the triangle for the 10 N and 4 N forces considered above. Both indicate a resultant of 9.4 N.

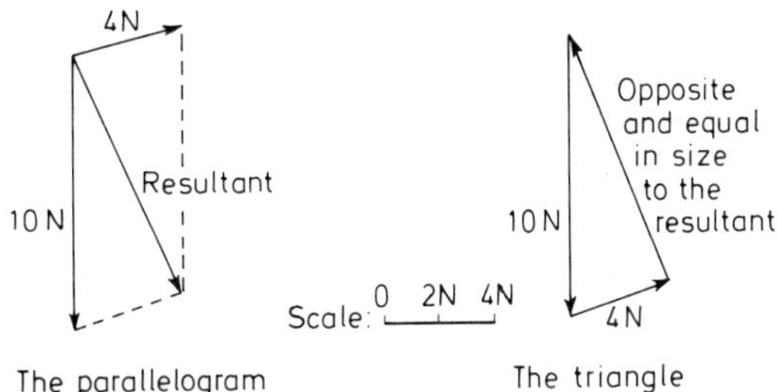

Figure 1.7

If more than two forces are involved then the polygon rule used for the determination of the equilibrium condition can be used, the resultant being the same size but in the opposite direction to the equilibrium force.

Example 3 Determine the resultant of the two forces shown acting on the object in Figure 1.8.

Lines of length 2.5 cm and 2 cm are drawn with an angle of 60° between them to represent the two forces. The parallelogram is then completed and diagonal drawn. The length of the diagonal is 3.8 cm and thus the force is 7.6 N acting in the direction of the diagonal.

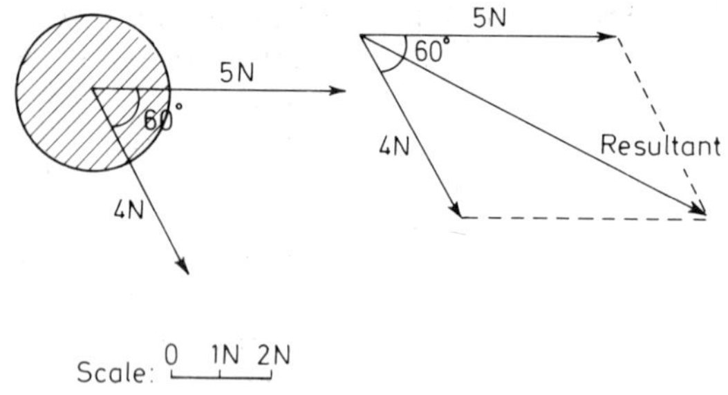

Figure 1.8

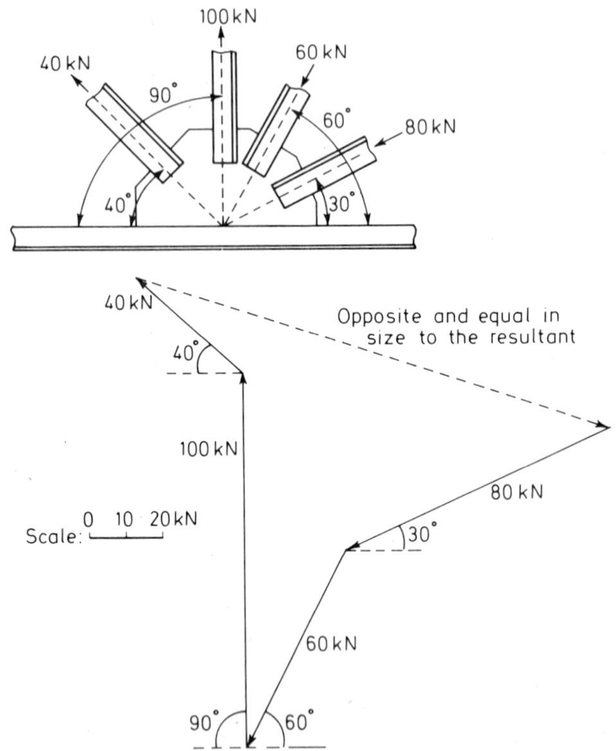

Figure 1.9

Example 4 Figure 1.9 shows a gusset plate acted on by forces along the four members linked by the gusset. What is the resultant force acting on the gusset plate due to these forces?

Lines are drawn, in Figure 1.9, to represent the size and direction of each force, the forces being taken in succession starting from the extreme right and working anticlockwise. If the forces had been in equilibrium then the force marked by the dotted line would have been needed to complete the polygon. Thus the resultant is the same size but in the opposite direction as the equilibrium force which completes the polygon. This has a length of 6.7 cm and so a force of 134 kN is indicated in the direction opposite to that of the dotted line.

RESOLVING FORCES

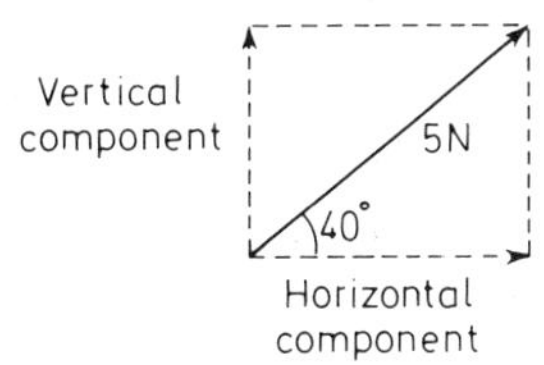

Figure 1.10

A resultant force can be determined for a system of forces, i.e. a number of forces can be replaced by just one force which has the same effect. The opposite of this process is to replace a single force by two forces at right angles to each other. This is known as *resolving* a force into its *components*. This is done by using the parallelogram law in reverse, i.e. starting with the resultant and ending with two forces.

Figure 1.10 shows a force of 5 N being resolved into two components at right angles to each other. A drawing can be used to find the components or they can be calculated. The horizontal component is $5 \cos 40^\circ = 3.8$ N (to two significant figures) and the vertical component is $5 \sin 40^\circ = 3.2$ N (to two significant figures).

The components of a force F at an angle of θ to the horizontal are

$$\text{Horizontal component} = F \cos \theta$$
$$\text{Vertical component} = F \sin \theta$$

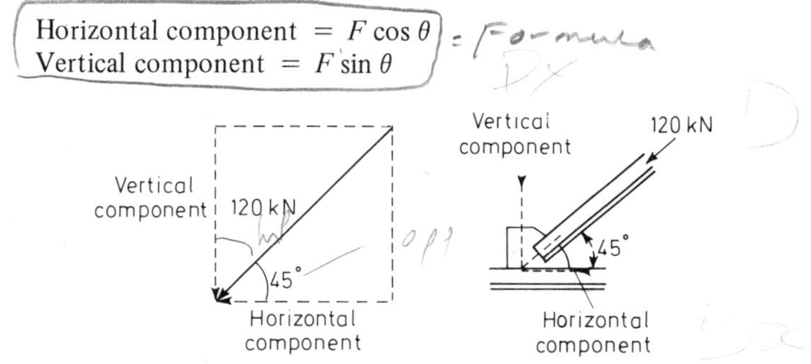

Figure 1.11

Example 5 Figure 1.11 shows a beam at an angle of 45° to the horizontal. What are forces acting in the horizontal and the vertical directions if the beam has a force of 120 kN acting along it?

Horizontal component $= 120 \cos 45^\circ = 85$ kN (to two significant figures).

Vertical component $= 120 \sin 45^\circ = 85$ kN (to two significant figures).

REACTIVE FORCES

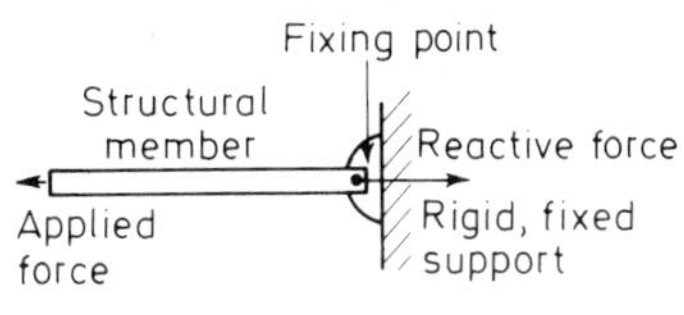

Figure 1.12

Figure 1.12 shows a structural member which is fixed at one end and subjected to an applied force at the other end. Since the member is in equilibrium there must be an equal and opposite force being exerted to balance the applied force and this is supplied by the fixing point of the member. It is known as the *reactive force*.

If the force acting on the member was 20 kN then the reactive force must be 20 kN in the opposite direction. As there are only these two forces acting on the member then, as it is in equilibrium, the lines of action of the two forces must be the same.

MOMENT OF A FORCE

Figure 1.13 shows a situation where the applied force results in a turning effect, the force causing the arm to rotate about its pivot point. The product of the force and the radius of its potential rotation about an axis is termed the *moment* of the force or *torque* about that axis. This is sometimes written as the product of the force F and the

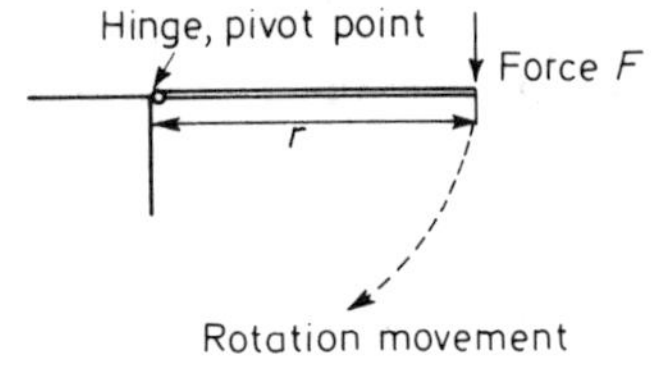

Figure 1.13

perpendicular distance of its line of action from the pivot axis r. The two definitions are identical in that both the distances are the same.

Moment of force F about A $= Fr$

UNIT: Newton metre (Nm).

If an object fails to rotate about a free pivot then the turning effect of one force must be balanced by the opposite direction turning effect of another force. In other words, the clockwise moments must equal the anticlockwise moments, both sets of moments being referred to the same pivot axis. Figure 1.14 shows a situation where the force F_1 a distance r_1 from the pivot would produce a potential clockwise rotation and the force F_2 a distance r_2 from the same pivot point would produce a potential anticlockwise rotation. If the object does not rotate then

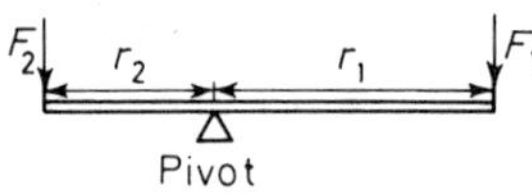

Figure 1.14

$$F_1 r_1 = F_2 r_2$$

Example 6 Calculate the force required at the right hand end of the pivoted beam in Figure 1.15 if the beam is to balance and not rotate.

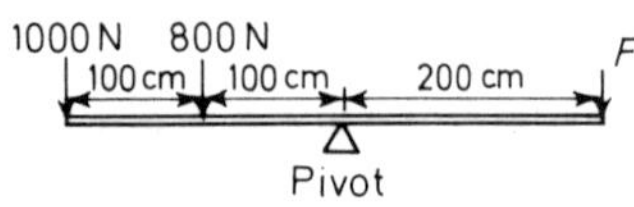

Figure 1.15

Considering the moments with reference to the pivot gives:

$$\text{Anticlockwise moments} = (1000 \times 200) + (800 \times 100) \text{ N cm}$$
$$= 280\,000 \text{ N cm}$$

This must equal the clockwise moment, thus

$$200F = 280\,000$$
$$F = 1400 \text{ N}$$

CONDITIONS FOR STATIC EQUILIBRIUM

An object is said to be in static equilibrium when there is no movement or tendency to movement in any direction. This requires that:

1. There must be no resultant force in any direction.
2. The sum of the anticlockwise moments about an axis equals the sum of the clockwise moments about the same axis.

The first of the above conditions can be written in terms of the components of the forces in two mutually perpendicular directions, i.e.,

Total of the upward components of forces = total of downward components of forces

Total of rightward components of forces = total of leftward components of forces.

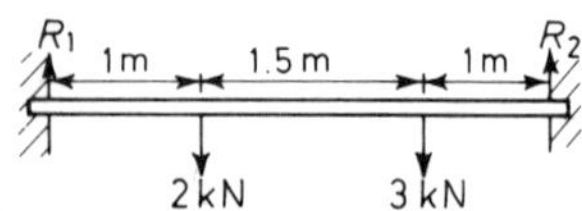

Figure 1.16

Example 7 Figure 1.16 shows a beam built into a wall at each end. The beam carries loads at the positions shown. What are the reactive forces at the walls?

As the beam is assumed to be in equilibrium the total of upwards forces must equal the total of downward forces. The reactive forces must be vertical forces if they are to balance the downward forces. Thus

$$R_1 + R_2 = 2 + 3 = 5 \text{ kN}$$

There are no sideways forces acting on the beam.

Taking moments about the end where reactive force R_1 acts gives

$$\text{Clockwise moments} = (2 \times 1) + (3 \times 2.5) = 9.5 \text{ kN m}$$

$$\text{Anticlockwise moments} = R_2 \times 3.5$$

As the anticlockwise and the clockwise moments must be equal then

$$3.5R_2 = 9.5$$

$$R_2 = 2.7 \text{ kN to two significant figures.}$$

But $R_1 + R_2 = 5$, hence $R_1 = 2.3$ kN.

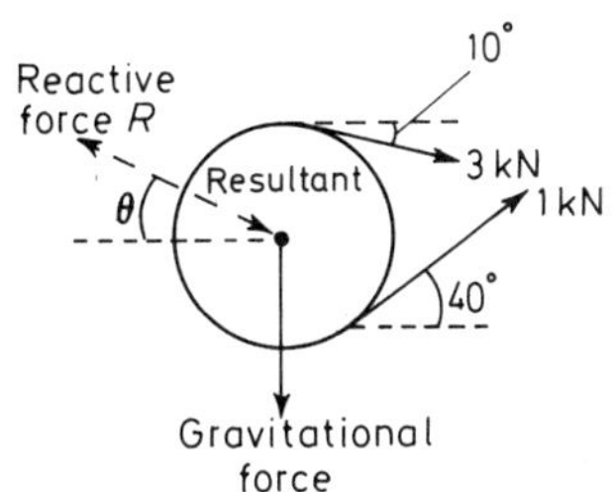

Figure 1.17

Example 8 Figure 1.17 shows a rope passing round a pulley of mass 20 kg. The tensions in the tight and slack sides of the rope are 3 kN and 1 kN. Calculate the resultant force acting on the pulley.

Suppose the resultant force, of magnitude R, be in the direction shown. Then if the pulley is in equilibrium, there must be a reactive force equal and opposite to resultant force.

$$\text{Upward component of the forces} = 1000 \sin 40^\circ + R \sin \theta$$

The force R referred to is the reactive force.

$$\text{Downward components of the forces} = 3000 \sin 10^\circ + 20 \times 9.8$$

That last term is the gravitational force acting on a mass of 20 kg. As the upward components must equal the downward components then

$$1000 \sin 40^\circ + R \sin \theta = 3000 \sin 10^\circ + 20 \times 9.8$$

$$R \sin \theta = 521 + 196 - 642$$

$$= 75 \text{ N}$$

The leftward components of the forces must equal the rightward components of the forces, thus

$$R \cos \theta = 3000 \cos 10^\circ + 1000 \cos 40^\circ$$

$$= 2954 + 766$$

$$= 3720 \text{ N}$$

Dividing the two equations involving R and θ gives

$$\frac{R \sin \theta}{R \cos \theta} = \frac{75}{3720}$$

$$\tan \theta = 0.0202$$

$$\theta = 1^\circ\, 14'$$

As $R \sin \theta = 75$ then substituting this value of θ gives

$R = 3500$ N to two significant figures.

CENTRE OF MASS

If you try to balance a ruler across a pencil there is just one position of the ruler for which balance occurs. This is when the anticlockwise moment of the ruler material to the left of the pivot axis balances the clockwise moment of the ruler material to the right of the pivot. The pivot axis at which this balance occurs passes through a point called the *centre of mass*. If you try to balance any object, whatever its form, then when balance occurs the clockwise moments are equal to the anti-clockwise moments and the balance axis passes through the centre of mass. The term *centre of gravity* is sometimes used instead of centre of mass.

When the rule is balanced across the pencil, the centre of mass of the rule is directly over the pencil, i.e. the pivot. The term 'centre of mass' is used because the total mass is acting at that one point, i.e. directly over the pencil. The total moment of the ruler is the same as that of a mass, equal to the total ruler mass, placed at the centre of mass.

Thus if the beam is considered to be made of a large number of segments, each having mass and hence weight,

segment no. 1 weight δw_1 a distance x_1 from the centre of mass

segment no. 2 weight δw_2 a distance x_2 from the centre of mass

segment no. 3 weight δw_3 a distance x_3 from the centre of mass

then the total moment of all these segments is

$$\delta w_1 x_1 + \delta w_2 x_2 + \delta w_3 x_3 + \text{ etc.}$$

This can be represented as

$$\text{total moment} = \text{sum of all } \delta wx \text{ terms.}$$

or

$$\text{total moment} = \Sigma \delta wx$$

The symbol Σ stands for 'sum of'. If we are replacing the entire beam by a concentrated load at just one point then this load W and at a distance $\bar{x}$ away must satisfy the condition

$$W\bar{x} = \Sigma \delta wx$$

$$\bar{x} = \frac{\Sigma \delta wx}{W}$$

$\bar{x}$ is the distance of the centre of mass from the point about which we are considering the moments. W is the total weight of the beam and must therefore be equal to the sum of the weights of all the segments.

If we take moments about the centre of mass, i.e. the balance point, then the anticlockwise moments equal the clockwise moments, i.e. the sum of all the δwx terms is zero. This means $\bar{x}$ will be zero and

$$\Sigma \delta wx = 0$$

For symmetrical homogeneous objects the centre of mass is the geometrical centre. Thus, for a sphere the centre of mass is at the centre. The sphere can thus be treated as an object with all its mass concentrated and acting at its centre. For a cube, the centre of mass is the centre of the cube. Again all the mass of the cube may be considered to be acting at its centre.

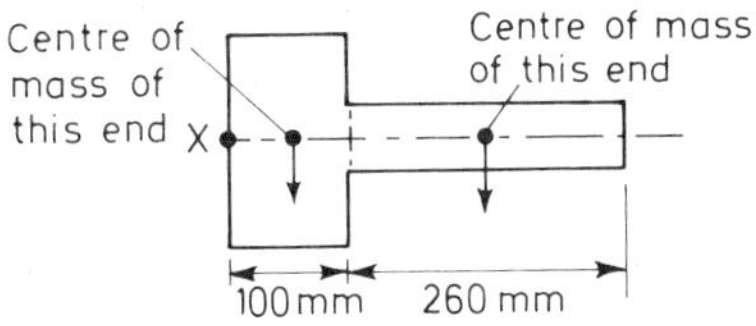

For composite objects, the centre of mass can be determined by considering the object to be made up of a number of smaller objects each with its mass concentrated at its own centre of mass. Thus for the object shown in Figure 1.18 the two parts are two different sized rectangular pieces. The centre of mass of each piece is at its centre. Thus one piece has its centre of mass a distance along the centre line from X of 50 mm and the other a distance of 230 mm from X. If the first piece has a weight of 6 N and the second piece a weight of 4 N then taking moments about X gives for the moments total

Figure 1.18 Centre of mass problem

$$(6 \times 50) + (4 \times 230) = 1220 \text{ N mm}$$

We could have considered all the mass to have been located at the centre of mass position a distance $\bar{x}$ from X, then as the total weight is 10 N

$$10\bar{x} = 1220$$

$$\bar{x} = 122 \text{ mm}$$

The centre of mass of the composite object is 122 mm from X along the centre line.

CENTROIDS

The *centroid* is the centre of area, the point about which the area of a section is evenly distributed. If an object is of uniform thickness and homogeneous, i.e. the mass per unit area of the surface is constant, then the centre of mass lies along a line through the centroid. Whereas with the centre of mass the moments considered were the product of weight and distance, with the centre of area, the centroid, we consider the product of the area and its distance away from the point about which the moment is taken is considered. The moment is a moment of area, often termed the *first moment of area.*

moment of force = force size × distance to line of action of force

moment of area = area size × distance to centre of area

With a large area the total moment of area is the sum of the component moments of area about the same axis. Thus if a surface is considered to be made up of a large number of area segments,

segment no. 1 area δa_1 a distance y_1 from the centroid

segment no. 2 area δa_2 a distance y_2 from the centroid

segment no. 3 area δa_3 a distance y_3 from the centroid

etc.

then the total area moment of all these segments is:

$$\delta a_1 y_1 + \delta a_2 y_2 + \delta a_3 y_3 + \text{etc.}$$

This can be represented as

Total area moment = sum of all δay terms,

or

$$\text{Total area moment} = \Sigma \delta a y$$

The centroid is the centre area and can be considered as an area A at a distance $\bar{y}$ away, where

$$A\bar{y} = \Sigma\delta ay$$

$$\bar{y} = \frac{\Sigma\delta ay}{A}$$

$\bar{y}$ is the distance of the centre of area from the point about which we are considering the moments. A is the total area of the surface and must therefore be equal to the sum of the areas of all the segments.

If when moments of the area components are being taken the point about which they are taken is the centroid, then the area moments will balance. As $\bar{y}$ will be zero this means

$$\Sigma\delta ay = 0$$

For symmetrical shapes the centroid is its centre. Thus for a circle the centroid is its centre. For a square the centroid is the centre of the square. For a rectangle the centroid is the centre of the rectangle.

The position of the centroid of the rectangle can be found by drawing lines to bisect each of the sides. The centroid is where the lines meet. This geometrical construction gives the same answer as considering the rectangle to be composed of a large number of area strips and performing the summation of all the area moments before dividing by the total area. Figure 1.19 shows the geometrical construction.

Figure 1.19

For composite shapes the centroid can be determined by considering the shape to be made up of a number of smaller shapes each with its area effectively located at its centroid. Thus, for the shape shown in Figure 1.20, the two parts are two different size rectangles. The centroid of each part is at its centre. Thus area A_1 has its centroid a distance of 75 mm along the centre line from the point X and area A_2 has its centroid a distance of 350 mm along the same line from point X. Taking the area moments about the point X gives for the moment total

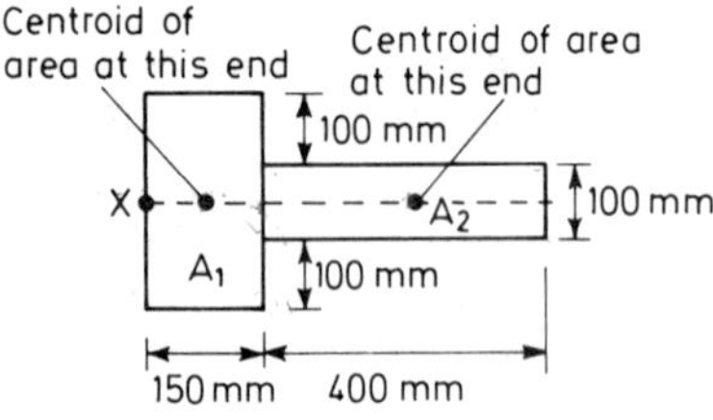

Figure 1.20 Centroid problem

$$75A_1 + 350A_2 = (75 \times 150 \times 300) + (350 \times 100 \times 400) \text{ mm} \times \text{mm} \times \text{mm}$$

$$= 17\,375\,000 \text{ mm}^3$$

We could have considered all the area to have been located at a point a distance $\bar{y}$ from X. The total area is $150 \times 300 + 100 \times 400$. Thus

$$\bar{y} = \frac{17\,375\,000}{(150 \times 300) + (100 \times 400)}$$

$$= 204 \text{ mm to three significant figures}$$

If the shape in Figure 1.20 had been the top of an object of constant depth and homogeneous then the centroid would have been directly over the centre of mass (Figure 1.21).

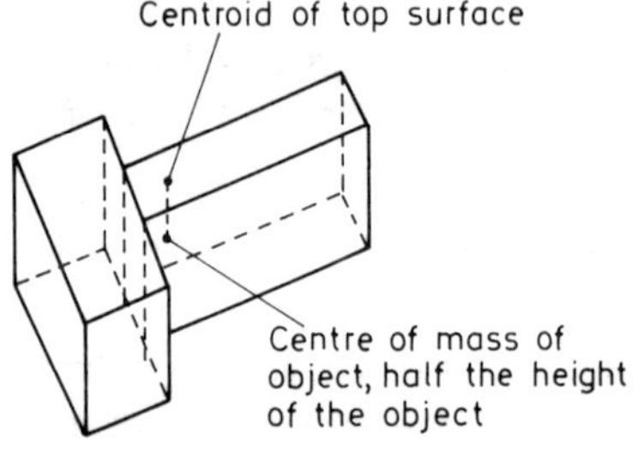

Figure 1.21

Example 9 Determine the centroid of the shape shown in Figure 1.22.

The centroid of the rectangle is at the centre, a distance of 0.25 m from point X along the centre line.

The centroid of the triangle can be found by drawing lines which bisect the sides. Where the lines meet gives the centroid.

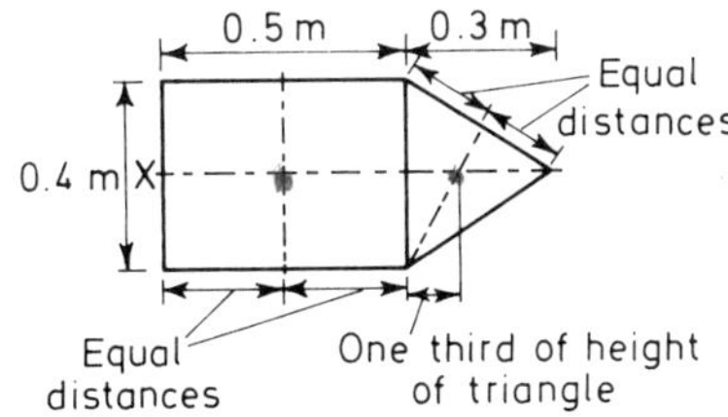

Figure 1.22

This is one third of the height of the triangle, i.e., 0.1 m above the base or 0.6 m from X.

Taking area moments about X gives for the moment total for the two shapes

$$(0.25 \times 0.5 \times 0.4) + (0.6 \times 0.5 \times 0.4 \times 0.3) \quad \text{m} \times \text{m} \times \text{m}$$

$$= 0.086 \text{ m}^3$$

The area of the triangle is half the base multiplied by the height.

The total area of the shape is $0.5 \times 0.4 + 0.5 \times 0.4 \times 0.3 = 0.26 \text{ m}^2$. Thus

$$\bar{y} = \frac{0.086}{0.26} = 0.33 \text{ m to two significant figures.}$$

If the shape had been that of the surface of a sheet of metal then the sheet could have been pivoted about a point 0.33 m from X along the centre line and would have balanced. The centre of mass of the sheet would have been half the thickness of the sheet directly under the centroid on the top surface.

PROBLEMS

1. What is the value of a 1500 N force in units of (a) kN, (b) MN?

2. Distinguish between vector and scalar quantities.

3. Explain what is meant by equilibrium.

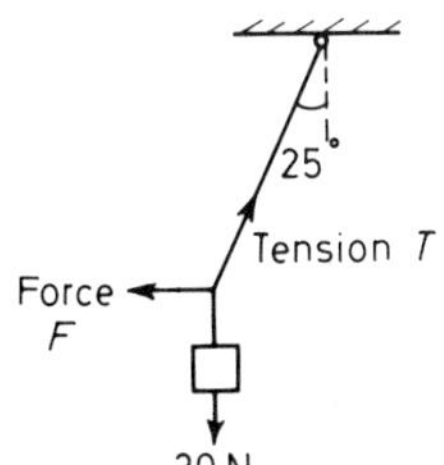

Figure 1.23

4. Determine the magnitude of the force required to maintain equilibrium for the suspended object shown in Figure 1.23 and the tension in the wire.

5. Determine the tensions in the cables used to support the object illustrated in Figure 1.24.

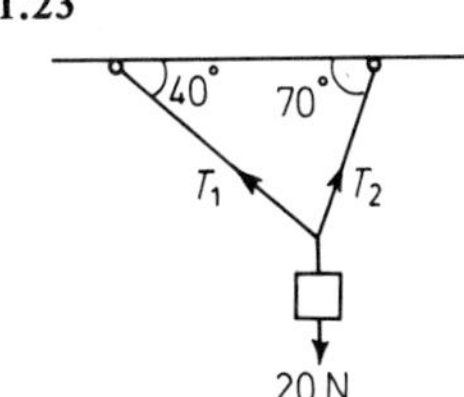

Figure 1.24

6. Determine the forces F_1 and F_2 in Figure 1.25, a simple jib.

7. Determine the size and the direction of the force needed to produce equilibrium for the force system shown in Figure 1.26.

8. Determine the resultant force when two forces of 20 kN and 40 kN act on an object if the angle between the lines of action of the forces is 90°.

9. What is the resultant force exerted on the gusset plate in Figure 1.27 by the forces shown.

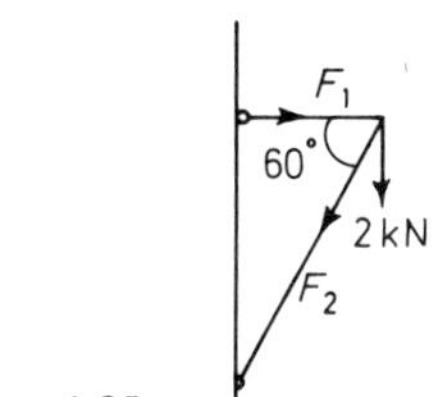

Figure 1.25

10. An object is being pulled along the floor by two men. If the forces applied to the object and the directions in which they are applied are as in Figure 1.28 what will be the resultant force acting on the object?

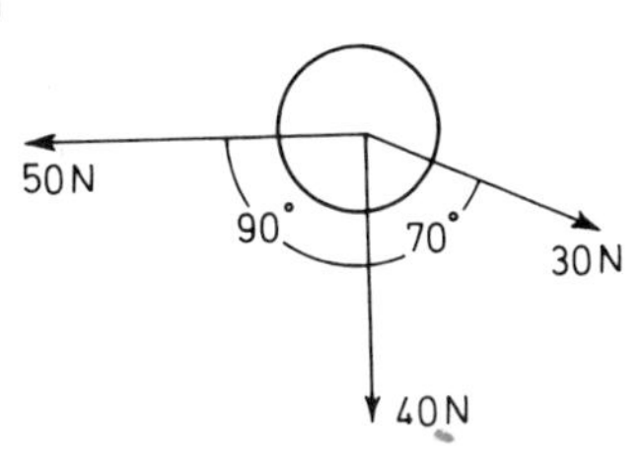

Figure 1.26

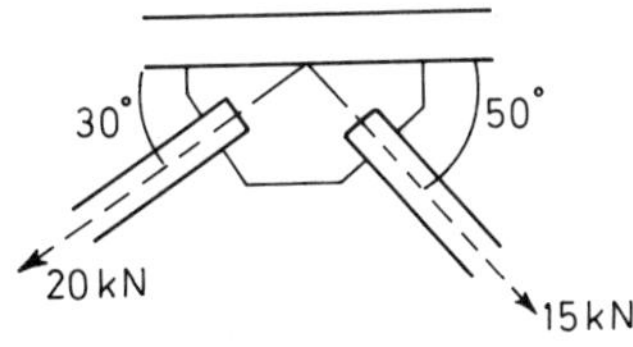

Figure 1.27

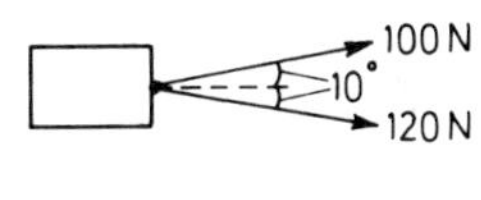

Figure 1.28

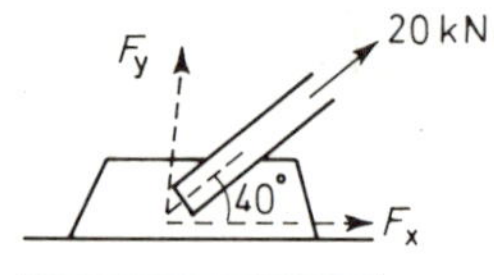

Figure 1.29

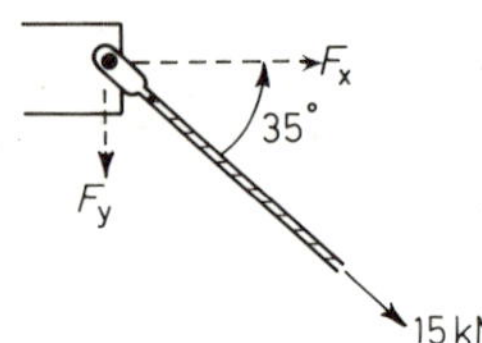

Figure 1.30

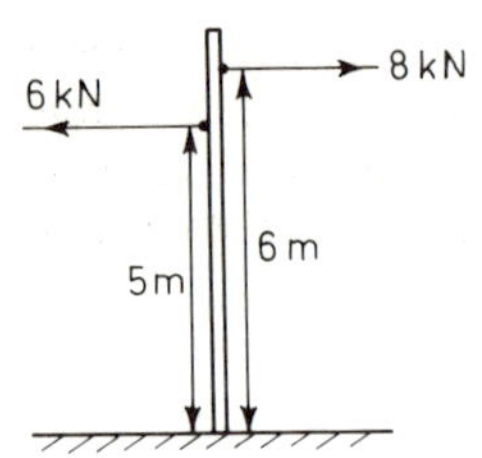

Figure 1.31

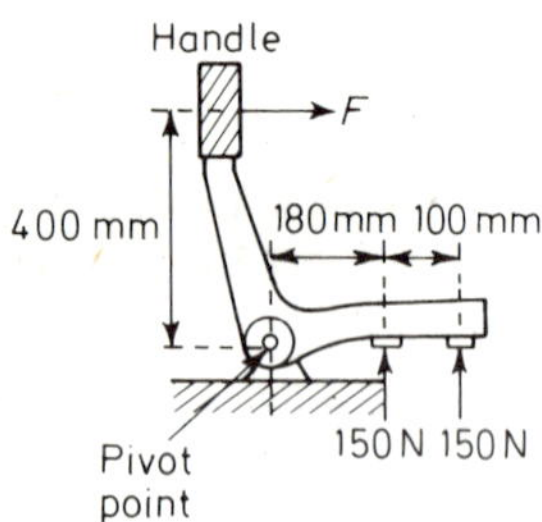

Figure 1.32

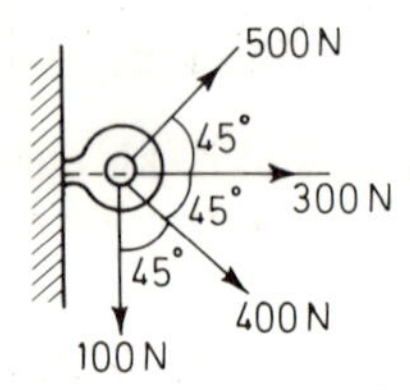

Figure 1.33

11. A structural member is acted on by a force of 20 kN in the direction shown in Figure 1.29. What are the components of that force in the x and y directions?

12. A cable exerts a force of 15 kN on a bracket (Figure 1.30). If the cable is at an angle of 35° to the horizontal what are the horizontal and vertical components of that force?

13. What is the reactive force at the pivot for the beam in Figure 1.15?

14. What is the reactive force at the bracket in Figure 1.30?

15. A taut horizontal cable of length 2 m has a mass of 10 kg suspended from its midpoint. What is the tension in the cable if the midpoint sags by 20 mm due to the mass being placed on it?

16. Masses of 2 kg and 4 kg are attached to opposite ends of a 3 m long beam. At what point along the beam should a support be placed for the beam to be in equilibrium? What will be the reactive force at the support point?

17. A beam 10 m long is supported at its ends. Loads of 2 kN and 3 kN are placed 2 m and 3 m respectively from one end. What are the reactive forces at the ends of the beam?

18. A mast has two cables attached to it. If the cables are positioned as shown in Figure 1.31 and the tensions in the cables are 8 kN and 6 kN what is the reactive force at the ground? What is the moment tending to cause the mast to rotate?

19. Figure 1.32 shows a control lever. Determine the size of the force F which will result in the resultant of the two 150 N forces and the force passing through the pivot point.

20. Determine the resultant force acting on the eyebolt shown in Figure 1.33 as a result of the four forces shown.

21. Explain the terms centre of mass and centroid.

22. Determine the positions of the centroids of the shapes shown in Figure 1.34.

23. A rod has a length of 200 mm and a uniform cross-section of 20 mm × 30 mm. Where is the centre of mass of the rod?

24. A rod has a length of 1 m. The first 500 mm of the rod has a cross-section of 30 mm × 30 mm and the second 500 mm a cross-section of 10 mm × 10 mm. If the material is of constant density where will the centre of mass be located?

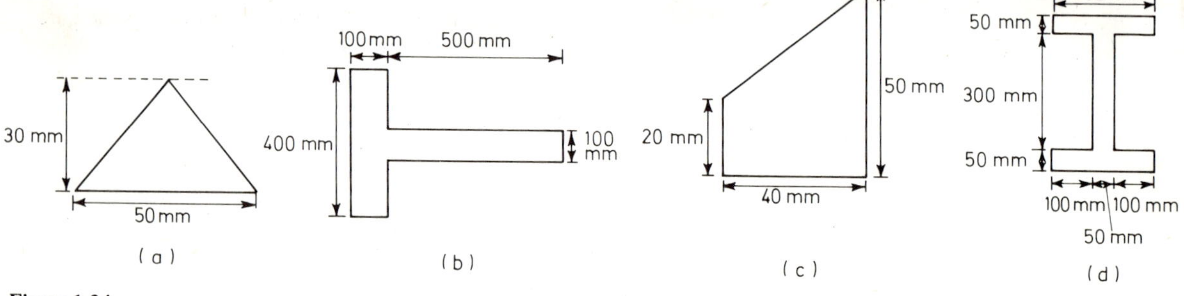

Figure 1.34

2 Stress and strain

STRESS

If a rubber strip is pulled, it stretches, and the more it is pulled, the more it stretches. There is a relationship between the extension and the applied force. This relationship can be expressed as a force/extension graph. Figure 2.1(a) shows the type of graph that might occur for a strip of rubber, Figure 2.1(b) the graph for a piece of steel. As will be apparent from the graphs the forces involved in the stretching of the steel are considerably greater than those with the rubber. The extensions produced with the rubber are considerably greater than those for the steel. The shapes of the graphs are also different. The rubber after a small force required very little increase in force to produce considerably greater extensions. After an extension of about 200 mm the rubber becomes much more difficult to stretch. The steel, however, requires quite large forces to produce any extension. However, when the force reaches 0.2 MN the steel increases considerably in length without any more force being applied. The steel is said to have yielded.

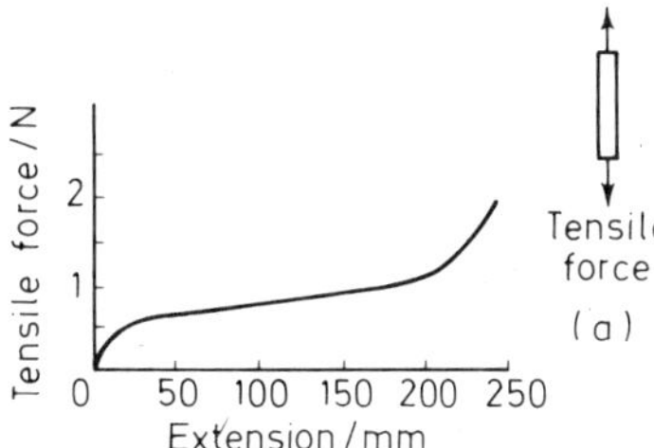

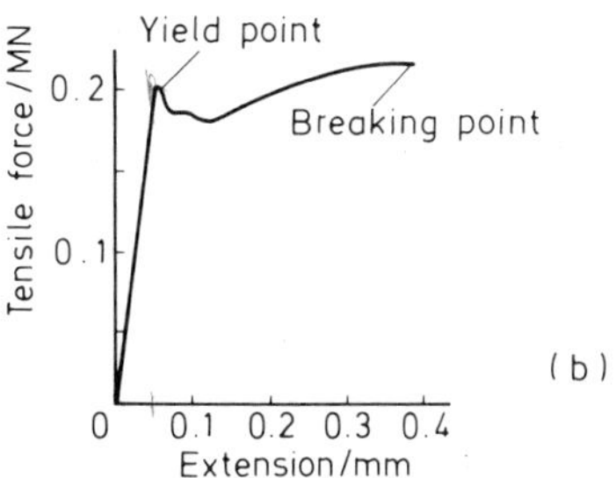

Figure 2.1 (a) Load/extension graph for a rubber strip. (b) Load/extension graph for mild steel

The forces applied to the rubber and the steel strips in Figure 2.1 produced an increase in the lengths of the strips. The forces are said to be *tensile forces* and the strips are in tension.

If the cross-sectional area of the strip being pulled is doubled then twice the force is needed to give the same extension. If the cross-sectional area is three times larger then the force has to be three times larger for the same extension. For the same extension the force per unit area has to be constant. This quantity is called *stress* σ.

$$\text{Stress} = \frac{\text{force}}{\text{area}}$$

UNIT: N/m^2 or Pa (1 pascal = 1 N/m^2)

The stress is said to be a *direct stress* when the area being stressed is at right angles to the line of action of the force. This is the case with the tensile force shown in Figure 2.1(a).

Example 1 A specimen having a cross-sectional area of 50 mm^2 was subjected to a tensile force of 0.1 MN. Calculate the stress induced in the specimen.

$$\text{Stress} = \frac{\text{force}}{\text{area}} = \frac{0.1}{50} \quad \frac{\text{MN}}{\text{mm}^2}$$

$$= 0.002\ \text{MN/mm}^2$$

$$= 2\ \text{kN/mm}^2$$

$$= 2\ \text{GN/m}^2 = 2\text{GPa}$$

1 MN = 10^6 N; 1 GN = 10^9 N; 1 kN = 10^3 N.
1 GN/m^2 = 1 kN/mm^2 = 1 GPa.

Example 2 A circular cross-section solid strut has a radius of 25 mm and is acted on by a tensile force of 20 kN. What is the stress acting on the strut?

$$\text{Stress} = \frac{\text{force}}{\text{area}} = \frac{20}{\pi 25^2} \quad \frac{\text{kN}}{\text{mm}^2}$$

Area $= \pi r^2$, where r is the radius.

Thus the stress is 0.010 kN/mm² or 10 N/mm² or 0.010 GPa, to two significant figures.

Example 3 A pipe has an outside diameter of 50 mm and an inside diameter of 45 mm and is acted on by a tensile force of 50 kN. What is the stress acting on the pipe?

$$\text{Cross-sectional area} = \frac{\pi}{4}(D^2 - d^2)$$

$$= \frac{\pi}{4}(50^2 - 45^2)$$

$$= 370 \text{ mm}^2 \text{ to two significant figures.}$$

Hence

$$\text{Stress} = \frac{\text{force}}{\text{area}} = \frac{50}{370} \quad \frac{\text{kN}}{\text{mm}^2}$$

$$= 0.14 \text{ kN/mm}^2 \text{ to two significant figures.}$$

STRAIN

If the strip being pulled by tensile forces is doubled in length then for the same stress the extension is doubled. If the strip is three times as long then for the same stress the extension is three times greater. For the same stress the same extension per unit length is produced. This quantity is called *strain* (symbol ϵ).

$$\text{Strain} = \frac{\text{change in length}}{\text{original length}}$$

There are no units for strain as it is a ratio of quantities with the same units, and is sometimes represented as a percentage.

Example 4 A strip of rubber of length 100 mm is acted on by a tensile force and extends by 2 mm. What is the strain?

$$\text{Strain} = \frac{\text{change in length}}{\text{original length}} = \frac{2}{100} \quad \frac{\text{mm}}{\text{mm}}$$

$$= 0.02$$

As a percentage, the strain is

$$\frac{2}{100} \times 100\% = 2\%$$

Example 5 A steel tensile test specimen has a gauge length of 50 mm. This increases by 0.030 mm when subject to tensile forces. What is the strain?

$$\text{Strain} = \frac{\text{change in length}}{\text{original length}} = \frac{0.030}{50} \quad \frac{\text{mm}}{\text{mm}}$$

$$= 0.0060 \text{ or } 0.6\%$$

MODULUS OF ELASTICITY

For the early part of the graph shown in Figure 2.1(b) the extension is directly proportional to the force. For that region of the graph the material is said to obey *Hooke's law*. This also means that the strain is directly proportional to the stress.

$$\text{Strain} \propto \text{stress}$$

or

$$\frac{\text{Stress}}{\text{Strain}} = \text{a constant}, E$$

The constant of proportionality E is called the modulus of elasticity or Young's modulus, the units of which are N/m^2 (generally kN/mm^2).

The greater the value of Young's modulus the greater the stress required for a particular strain, i.e. the greater the force for a particular extension. Thus a material with a high Young's modulus is more difficult to extend than material with a lower Young's modulus (Figure 2.2). Young's modulus can be considered as a measure of the stiffness of a material, the higher the modulus the stiffer the material.

The following are typical values of Young's modulus.

$1 \text{ kN/mm}^2 = 10^9 \text{ N/m}^2 = 10^9 \text{ Pa} = 1 \text{ GPa}$.

Material	*Young's modulus*/kN mm^{-2}
Mild steel	210
Cast iron	150
Brass	120
Copper	120
Aluminium alloys	70
Concrete	20
Polythene (high density)	1.4
Rubber	0.007

Higher Young's modulus, harder to stretch
Lower Young's modulus, easier to stretch
Force
0
Extension
(a)

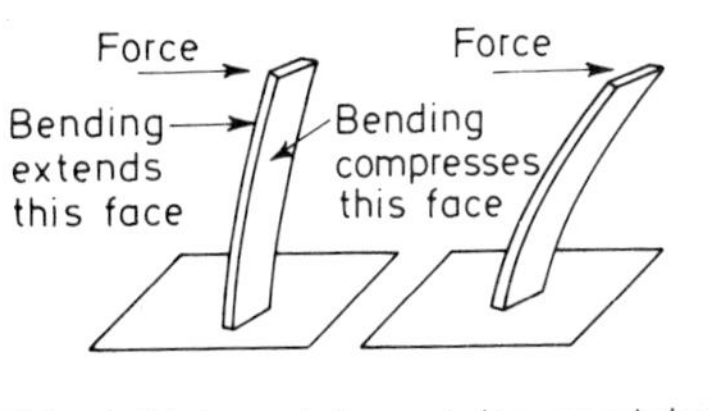

Figure 2.2 (a) The higher the Young's modulus, the harder it is to stretch a material. (b) The same force applied to identical size bars results in the greater bending in the bar of low Young's modulus

Where the graph of extension against force, or strain against stress, is not a straight line passing through the origin for the early stages of the graph, i.e. Hooke's law is not obeyed, then the value quoted for Young's modulus represents the straight line that the graph best approximates to. Thus in the case of rubber, Figure 2.1(a), the Young's modulus represents the value obtained for the straight line that the early part of the graph most closely resembles.

Example 6 Calculate the strain that will be produced by a tensile stress of 10 N/mm^2 extending a bar of aluminium alloy with a Young's modulus of 70 kN/mm^2.

$$\frac{\text{Stress}}{\text{Strain}} = E$$

Hence

$$\text{Strain} = \frac{\text{Stress}}{E} = \frac{10}{70} \qquad \frac{\text{N/mm}^2}{\text{kN/mm}^2}$$

$$= \frac{10}{70 \times 10^3} \qquad \frac{\text{N/mm}^2}{\text{N/mm}^2}$$

$= 0.0014$ or 0.14% to two significant figures.

Example 7 Calculate the strain that will be produced by a tensile stress of 10 N/mm^2 acting on a bar of mild steel with a Young's modulus of 210 kN/mm^2.

$$\text{Strain} = \frac{\text{Stress}}{E} = \frac{10}{210} \qquad \frac{\text{N/mm}^2}{\text{kN/mm}^2}$$

$$= \frac{10}{210 \times 10^3} \qquad \frac{\text{N/mm}^2}{\text{N/mm}^2}$$

$= 0.000\,048$ or 0.0048% to two significant figures.

This is the same stress as used with the aluminium alloy in the previous question but the strain produced is considerably smaller. The mild steel is 'stiffer' than the aluminium alloy, not so easily stretched.

Example 8 A tie bar has two holes a distance of 4 m apart. By how much does this distance increase when a tensile load of 20 kN is applied to the tie bar? The tie bar is a rectangular section 40 mm × 10 mm and the material of which the tie bar is made has a modulus of elasticity of 210 kN/mm^2.

$$\text{Stress} = \frac{\text{force}}{\text{area}} = \frac{20}{40 \times 10} \qquad \frac{\text{kN}}{\text{mm} \times \text{mm}}$$

$$\text{Strain} = \frac{\text{extension}}{\text{original length}} = \frac{x}{4000}$$

But

$$\text{Strain} = \frac{\text{stress}}{E} = \frac{20}{40 \times 10 \times 210} \qquad \frac{\text{kN/mm}^2}{\text{kN/mm}^2}$$

Hence

$x = 0.95$ mm to two significant figures.

YIELD STRESS

The load/extension graph for mild steel in Figure 2.1(b) shows a sharp discontinuity when the material suddenly yields and little or no force is needed to cause a quite significant increase in length. This *yield point* is close to or coincident with the limit of proportionality.

$$\text{Yield stress} = \frac{\text{load at the yield point}}{\text{original cross-section area}}$$

The units of yield stress are N/m^2 or Pa.

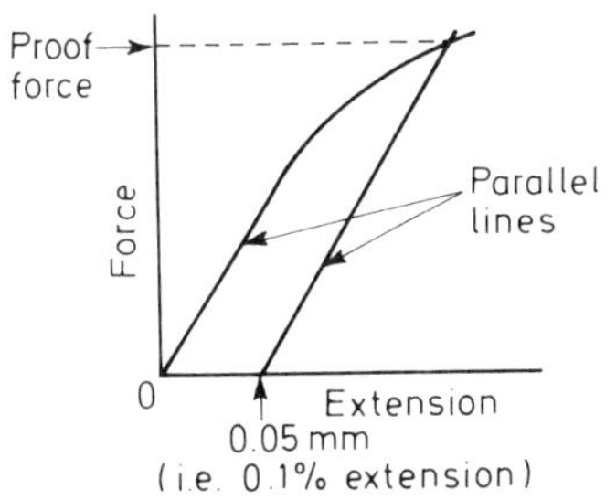

Figure 2.3

Some materials have no pronounced yield point and instead of quoting the yield stress for such materials it is customary to quote the *proof stress*. Figure 2.3 shows a force/extension graph for such a material. The proof force is determined from the graph by drawing a line parallel to the linear part of the force/extension graph but displaced by an extension equivalent to a strain of 0.1% in the case where the 0.1% proof stress is required. Thus for a gauge length of 50 mm this means a displacement of 0.05 mm from the straight line force/extension graph. Where this parallel line occurs the force/extension graph gives the proof force. Hence the proof stress can be calculated. A similar method can be used where the 0.2% proof stress is required.

The following are some typical values of yield stress and 0.1% proof stress.

Material	*Stress*/N mm^{-2}
Mild steel	Yield stress 200
Aluminium alloy	0.1% proof stress 157
Brass	0.1% proof stress 280
Copper	0.1% proof stress 300

Example 9 A test piece has a diameter of 5.64 mm and is found to yield under a force of 5.2 kN. What is the yield stress?

$$\text{Yield stress} = \frac{\text{load at yield point}}{\text{original cross-section area}}$$

$$= \frac{5.2}{\tfrac{1}{4}\pi\, 5.64^2} \quad \frac{\text{kN}}{\text{mm}^2}$$

$$= 208 \text{ N/mm}^2 \text{ to three significant figures}$$

Example 10 What is the 0.1% proof force for a specimen of diameter 5.64 mm if the 0.1% proof stress is 300 N/mm^2?

$$0.1\% \text{ proof stress} = \frac{0.1\% \text{ proof force}}{\text{original cross-section area}}$$

Hence

$$0.1\% \text{ proof force} = \frac{300}{\tfrac{1}{4}\pi\, 5.64^2} \quad \frac{\text{N/mm}^2}{\text{mm}^2}$$

$$= 12.0 \text{ N to three significant figures.}$$

STRENGTH

The *strength* of a material is the maximum stress the material can withstand. The following are typical strength values.

Material	*Tensile strength*/N mm^{-2}
Mild steel	400
Cast iron	140–300
Brass	120–400
Copper	140
Aluminium alloys	140–600
Concrete	5
Polythene (high density)	30
Rubber	30

Comparing the above table with that given earlier for Young's modulus it will be seen that a high value for Young's modulus does not necessarily mean a high value for the strength.

Example 11 Calculate the maximum force in tension a steel bar of cross-section 20 mm × 10 mm can withstand. The tensile strength of the material is 400 N/mm².

$$\text{Tensile strength} = \frac{\text{maximum force}}{\text{area}}$$

Hence

$$\begin{aligned}\text{Maximum force} &= \text{tensile strength} \times \text{area}\\ &= 400 \times 20 \times 10 \qquad \text{N/mm}^2 \times \text{mm}^2\\ &= 80\ \text{kN}\end{aligned}$$

Example 12 Calculate the maximum force in tension an aluminium alloy bar of cross-section 20 mm × 10 mm can withstand. The tensile strength of the material is 300 N/mm².

$$\begin{aligned}\text{Maximum force} &= \text{tensile strength} \times \text{area}\\ &= 300 \times 20 \times 10 \qquad \text{N/mm}^2 \times \text{mm}^2\\ &= 60\ \text{kN}\end{aligned}$$

FACTOR OF SAFETY

A strut in a girder bridge is expected to withstand stress without breaking. The material used for the body of a car is expected to withstand normal use without breaking. Thus the stress while in use is normally not expected to become so high that the material breaks. Because the stress that may be realized in practice is often not known, a factor of safety is often used. The ratio of the tensile strength, i.e. the maximum stress the material can withstand, to the maximum permissible stress in practice is known as the *factor of safety.*

$$\text{Factor of safety} = \frac{\text{tensile strength}}{\text{maximum permissible stress}}$$

The factor of safety has no units as it is a ratio.

Though a strut in a girder bridge is expected to withstand stress without breaking it is also expected to withstand stress without yielding and assuming a permanent deformation. The body of a car must not yield and permanently deform. It would be rather inconvenient if every time a car was driven it changed shape. Thus the stress such a material is subject to should not be so high that the yield stress is reached. There is thus generally greater use made of defining the factor of safety in terms of the yield stress.

$$\text{Factor of safety} = \frac{\text{yield stress}}{\text{maximum permissible stress}}$$

The factor of safety is sometimes called the 'factor of ignorance' because it makes allowances for all the things we do not know about the materials in use, as opposed to the behaviour found during tensile tests, and the stress applied in use rather than in theoretical calculations. For dead loads, a factor of safety of four or more is often used.

An alternative to the factor of safety, or ignorance, is a designation of safe working stresses for a particular material.

Example 12 A sample of steel has a yield stress of 200 N/mm^2, what should be the maximum permissible stress if a safety factor of 4 is used?

$$\text{Maximum permissible stress} = \frac{\text{yield stress}}{\text{factor of safety}}$$

$$= \frac{200}{4} \quad \text{N/mm}^2$$

$$= 50 \text{ N/mm}^2$$

COMPRESSION

When the forces applied to a piece of material squash that material, i.e. reduce the length in the direction of the force, then the forces are said to be *compressive* forces and the material is in compression (Figure 2.4). For compression we have

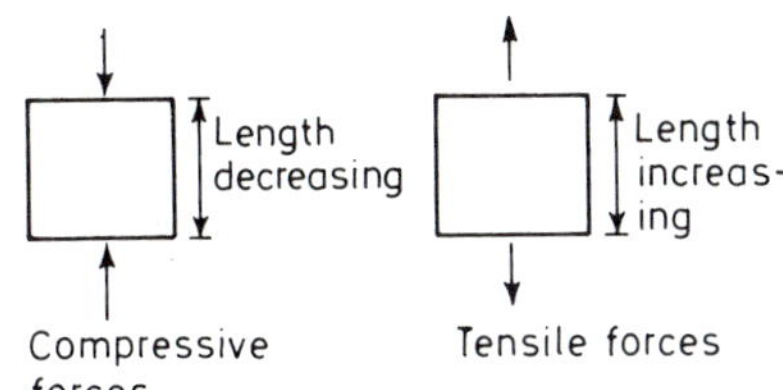

Figure 2.4

$$\text{Compressive stress} = \frac{\text{force}}{\text{area}}$$

the area concerned being that at right angles to the line of action of the force;

$$\text{Compressive strain} = \frac{\text{change in length}}{\text{original length}}$$

the change in length being a reduction in the original length;

$$\text{Modulus of elasticity} = \frac{\text{compressive stress}}{\text{compressive strain}}$$

for that part of the stress/strain graph for which the strain is directly proportional to the stress.

Metals, in general, have essentially the same modulus of elasticity in compression as in tension. The maximum stress a metal can withstand, i.e. the tensile strength, is however generally much smaller than the stress at which the material fails in compression.

Example 13 A strut of diameter 50 mm has a compressive load of 60 kN applied to it. Calculate the compressive stress.

$$\text{Compressive stress} = \frac{\text{force}}{\text{area}}$$

$$= \frac{60}{\dfrac{\pi 50^2}{4}} \quad \frac{\text{kN}}{\text{mm}^2}$$

$$= 0.031 \text{ kN/mm}^2 = 31 \text{ N/mm}^2 \text{ to two significant figures.}$$

Example 14 A pipe of external diameter 60 mm and internal diameter 50 mm is subject to a compressive force of 50 kN. What is the compressive stress?

$$\text{Compressive stress} = \frac{\text{force}}{\text{area}}$$

$$= \frac{50}{\frac{\pi}{4}(60^2 - 50^2)} \qquad \frac{\text{kN}}{\text{mm}^2}$$

$$= 0.058 \text{ kN/mm}^2 = 58 \text{ N/mm}^2 \text{ to two significant figures.}$$

Example 15 By how much does a bar of mild steel compress when subject to a compressive force of 60 kN if the bar is of rectangular cross-section 40 mm × 20 mm and length 5 m?

The modulus of elasticity is to be taken as 210 kN/mm².

$$\text{Compressive stress} = \frac{60}{40 \times 20} \qquad \frac{\text{kN}}{\text{mm}^2}$$

$$= 0.075 \text{ kN/mm}^2$$

$$\text{Compressive strain} = \frac{\text{change in length } x}{5000}$$

$$\text{Compressive strain} = \frac{\text{compressive stress}}{\text{modulus of elasticity}}$$

$$\frac{x}{5000} = \frac{0.075}{210} \qquad \frac{\text{kN/mm}^2}{\text{kN/mm}^2}$$

$x = 1.8$ mm to two significant figures.

Example 16 A machine is to be mounted on a rubber pad. The pad has to carry a load of 6 kN and have a maximum compression of 2 mm under this load. The maximum stress that is allowed for the rubber is 0.250 N/mm². What is the size of the pad that would be appropriate for these maximum conditions? The modulus of elasticity for the rubber can be taken as 5 N/mm².

$$\text{Compressive stress} = \frac{\text{force}}{\text{area}}$$

Thus

$$\frac{\text{N}}{\text{mm}^2} \quad 0.250 = \frac{6 \times 10^3}{\text{area}} \qquad \text{N}$$

$$\text{Area} = 24 \times 10^3 \text{ mm}^2$$

$$\text{Compressive strain} = \frac{\text{compressive stress}}{\text{modulus of elasticity}}$$

$$= \frac{0.250}{5} \qquad \frac{\text{N/mm}^2}{\text{N/mm}^2}$$

$$= 0.05$$

But

$$\text{Compressive strain} = \frac{\text{compression}}{\text{original length}}$$

Hence

$$\text{Original length} = \frac{2}{0.05} \text{ mm}$$
$$= 40 \text{ mm}$$

The pad would thus have an area of 24×10^3 mm^2 and a thickness of 40 mm .

STRESSES DUE TO CHANGES IN TEMPERATURE

For most materials an increase in temperature results in expansion, i.e. a bar of the material becomes longer. The amount by which the bar becomes longer, or shorter if the temperature drops, depends on the material concerned, the initial length of the material and the size of the temperature change.

Change in length $\propto$ original length
Change in length $\propto$ temperature change

Doubling the initial length doubles the change in length for a given temperature change. Doubling the temperature change doubles the change in length for a given initial length.

$$\text{Change in length} = L_t - L_0 = L_0 \alpha t$$

where L_0 = original length, L_t = length after a temperature change of t°, α = coefficient of linear expansion. The coefficient depends on the material concerned and has units of /°C, °C^{-1}, /K or K^{-1}. The table below gives typical values of coefficients.

Material	*Coefficient of linear expansion*/10^{-6} °C^{-1}
Mild steel	11
Cast iron	10
Brass	18
Copper	17
Aluminium alloys	23
Concrete	14

Thus, because brass has a higher coefficient of linear expansion than mild steel, a bar of brass will expand more than the same initial size bar of steel for the same temperature rise.

A bar of mild steel 2000 mm long will thus expand by $2000 \times 11 \times 10^{-6} \times 5$ for a 5 °C change in temperature. This amounts to 11×10^{-2} or 0.11 mm.

Suppose, however, that the bar of steel is rigidly attached to some other components and so is prevented from expanding. Though the temperature rises by 5 °C the bar is not able to freely expand and so does not increase in length by 0.11 mm. The strain produced as a result of this restriction is the same as if the bar had been compressed by 0.11 mm to fit within the 2000 mm distance at this higher temperature.

$$\text{Strain} = \frac{\text{change in length}}{\text{original length}}$$

Thus

$$\text{Strain} = \frac{L_t - L_0}{L_0}$$

$$\text{Strain} = \frac{L_0 \alpha t}{L_0} = \alpha t$$

This strain will give rise to a stress.

$$\text{Stress} = E \times \text{strain}$$

$$\text{Stress} = E\alpha t$$

This is the stress produced as a result of not allowing the bar to expand. It is a compressive stress.

The above assumes that extension of the bar is completely prevented. If only part of the extension is prevented then the stress developed is correspondingly reduced.

Example 17 A bar of mild steel is constrained between two rigid supports. Calculate the stress developed in the bar as a result of a temperature rise of 8 °C. The coefficient of linear expansion can be taken as $11 \times 10^{-6}/°C$ and the modulus of elasticity as 210 kN/mm².

$$\begin{aligned} \text{Stress} &= E\alpha t \\ &= 210 \times 11 \times 10^{-6} \times 8 \quad \text{kN/mm}^2 \times /°\text{C} \times °\text{C} \\ &= 0.018\ \text{kN/mm}^2 \text{ to two significant figures.} \end{aligned}$$

Example 18 Steel railway lines are welded together and are constrained so that they cannot expand or contract. If the rails are unstressed at 15 °C what will be the stress developed in the rails at 25 °C? The modulus of elasticity can be taken as 210 kN/mm² and the coefficient of linear expansion as $11 \times 10^{-6}/°C$.

$$\begin{aligned} \text{Stress} &= E\alpha t \\ &= 210 \times 11 \times 10^{-6} \times 10 \quad \text{kN/mm}^2 \times /°\text{C} \times °\text{C} \\ &= 0.023\ \text{kN/mm}^2 \text{ to two significant figures.} \end{aligned}$$

COMPOSITE BARS

The term composite bar is used to describe the situation where there are two or more elements fixed together. Figure 2.5 shows a common form of composite bar, a steel reinforcement bar in a sheath of concrete, i.e. a reinforced concrete pillar.

Each element in Figure 2.5 has a share of the applied force. The sum of the forces acting on each element equals the total applied force.

$$\text{Total force} = \text{force on 1} + \text{force on 2}$$

The stress acting on element 1 is the force acting on 1 divided by the cross-sectional area of that element.

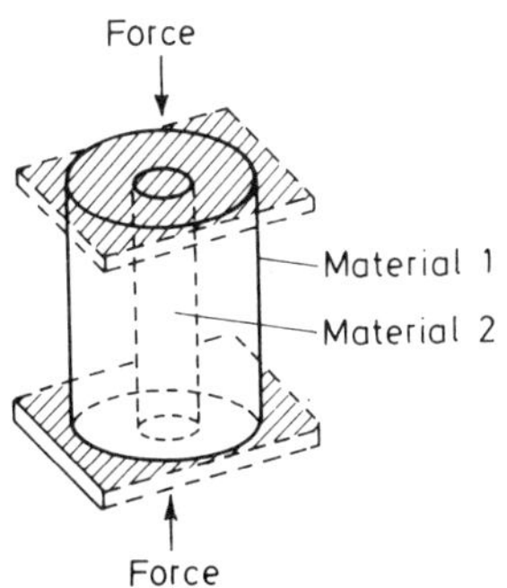

Figure 2.5 A composite bar with each element sharing the same load

$$\text{Stress on 1} = \frac{\text{force on 1}}{\text{area of 1}}$$

Similarly

$$\text{Stress on 2} = \frac{\text{force on 2}}{\text{area of 2}}$$

Hence

$$\text{Total force} = \text{stress on 1} \times \text{area of 1} + \text{stress on 2} \times \text{area of 2} \quad (2.1)$$

Both element 1 and element 2 must be compressed by the same amount and thus as the lengths of each element are the same then the strains on each element must be the same.

$$\text{Strain on 1} = \text{strain on 2}$$

But strain = stress/E, where E is the modulus of elasticity.
Thus

$$\frac{\text{stress on 1}}{E \text{ of 1}} = \frac{\text{stress on 2}}{E \text{ of 2}} \quad (2.2)$$

Both equations (2.1) and (2.2) above have to be satisfied if the two elements are to share the load and compress by the same amount. The stresses on each element can be obtained by solving these simultaneous equations.

Example 19 A reinforced concrete column of cross-sectional area 40 000 mm^2 includes a steel bar, of cross-sectional area 400 mm^2. What are the stresses carried by the concrete and the steel bar when the load on the column is 300 kN? The modulus of elasticity of the steel can be taken as 210 kN/mm^2 and that of the concrete as 20 kN/mm^2.

Total force = stress on 1 × area of 1 + stress on 2 × area of 2

$$300 = \sigma_1 \times 36\,000 + \sigma_2 \times 400$$

The stresses have units of kN/mm^2 as the force is in kN and the areas in mm^2.

$$\frac{\text{stress on 1}}{E \text{ of 1}} = \frac{\text{stress on 2}}{E \text{ of 2}}$$

$$\frac{\sigma_1}{20} = \frac{\sigma_2}{300}$$

The stresses have units of kN/mm^2 as the moduli have units of kN/mm^2.

The two simultaneous equations can be solved for σ_1 and σ_2.

$$\sigma_1 = \frac{20\sigma_2}{300}$$

Hence

$$300 = \frac{20\sigma_2}{300} \times 36\,000 + \sigma_2 \times 400$$

$$90\,000 = 720\,000\,\sigma_2 + 120\,000\,\sigma_2$$

$$\sigma_2 = \frac{90\,000}{840\,000}$$

$$\sigma_2 = 0.11 \text{ kN/mm}^2 \text{ to two significant figures.}$$

$$\sigma_1 = 0.0073 \text{ kN/mm}^2 \text{ to two significant figures.}$$

Example 20 A pile used for the foundation of a building consists of a cast iron cylinder filled with concrete. The cylinder has a diameter of 300 mm and a thickness of 25 mm. The load to be carried by the pile is 80 kN. What will be the stresses acting on the cast iron and the concrete? The modulus of elasticity of the cast iron is 150 kN/mm² and that of the concrete 20 kN/mm .

Total force = stress on 1 × area of 1 + stress on 2 × area of 2

$$80 = \sigma_1 \times \frac{\pi}{4}(300^2 - 275^2) + \sigma_2 \times \frac{\pi}{4} \times 275^2$$

$$320 = \sigma_1 \pi \times 14\,475 + \sigma_2 \pi \times 75\,525$$

The stresses have units of kN/mm² as the force is in kN and the areas in mm².

$$\sigma_1 = \frac{150}{20}\sigma_2$$

The two simultaneous equations can be solved for σ_1 and σ_2. Hence

$$320 = \frac{150}{20}\sigma_2 \pi \times 14\,475 + \sigma_2 \pi \times 75\,525$$

$$\sigma_2 = 0.000\,54 \text{ kN/mm}^2 \text{ to two significant figures}$$

or

$$\sigma_2 = 0.54 \text{ N/mm}^2.$$

$$\sigma_1 = 0.0041 \text{ kN/mm}^2 \text{ to two significant figures}$$

or

$$\sigma_1 = 4.1 \text{ N/mm}^2$$

Example 21 A brass bar of diameter 30 mm fits inside a cylindrical steel tube of internal diameter 35 mm and external diameter 40 mm. Calculate the axial load that can be applied to the composite bar if the stress in the brass is not to exceed 70 N/mm². The modulus of elasticity for the brass is 120 kN/mm² and that for the steel 210 kN/mm².

Total force = stress on 1 × area of 1 + stress on 2 × area of 2

$$F = 0.070 \times \frac{\pi}{4}30^2 + \sigma_2 \times \frac{\pi}{4}(40^2 - 35^2)$$

$$4F = 63\pi + 375\pi\sigma_2$$

The force will be in kN and the stress in kN/mm².

$$\frac{\text{stress on 1}}{E \text{ of } 1} = \frac{\text{stress on 2}}{E \text{ of } 1}$$

$$\frac{0.070}{120} = \frac{\sigma_2}{210}$$

Hence

$$\sigma_2 = 0.12 \text{ kN/mm}^2 \text{ to two significant figures.}$$

Substituting this value in the earlier equation gives:

$$4F = 63\pi + 375\pi \times 0.12$$

$$F = 86 \text{ kN to two significant figures.}$$

COMPOSITE BARS AND TEMPERATURE CHANGES

The amount by which a bar expands depends not only on its length and the temperature rise but also on the material concerned. Brass expands more than mild steel for the same initial length and the same temperature rise. Thus if bars of brass and steel were of equal length at some temperature and the temperature rises, the brass bar will become longer than the mild steel bar. If, however, the brass and the steel bars are fixed together so that the amount by which they extend must be the same for both then stresses are produced in the two materials as a result of any change in temperature.

Figure 2.6 illustrates the effect of temperature on a composite bar. Initially both the materials have the same length L_0. If material 1 had been free to expand then a temperature rise of t° would have resulted

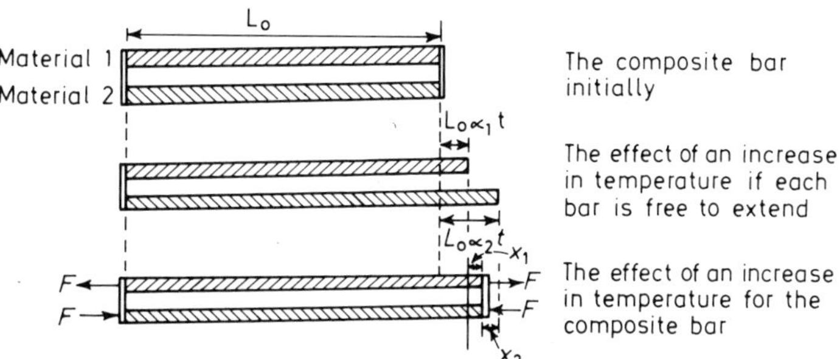

Figure 2.6 The effect on a composite bar of a rise in temperature. The expansion is exaggerated

in an expansion of $L_0\alpha_1 t$, where α_1 is the coefficient of expansion of material 1. If material 2 had been free to expand then a temperature rise of t° would have resulted in an expansion of $L_0\alpha_2 t$, where α_2 is the coefficient of expansion of material 2. Because the two materials are so constrained that they must both expand by the same amount the expansion of each differs from that they would have had if free. Material 1 is made to expand by more than $L_0\alpha_1 t$, the result is that material 1 is subject to tensile forces. Material 2 is made to expand less than $L_0\alpha_2 t$, the result is that material 2 is subject to compressive forces. Because the composite bar has no net force acting on it then the force acting at one end of material 1 must be opposite and equal to the force acting at the same end of material 2. The end fixing has thus to supply a force which causes an extra extension of material 1 and a compression of material 2. If the end fixing is not strong enough to withstand these forces then it will fracture.

As will be seen from Figure 2.5

$$L_0\alpha_2 t - L_0\alpha_1 t = x_2 + x_1 \tag{2.3}$$

where x_1 is the amount by which the force F causes material 1 to be extended and x_2 is the amount by which the force F causes material 2 to be compressed. The strain resulting for material 1 as a consequence of this force F is given by

$$\text{strain} = \frac{\text{stress}}{E}$$

$$\text{strain} = \frac{\sigma_1}{E_1}$$

where σ_1 is the stress produced in material 1 by the force F. The modulus of elasticity of material 1 is E_1. But the extension is x_1 for an initial length of $L_0 + L_0 a_1 t$. Thus

$$\text{strain} = \frac{x_1}{L_0 + L_0\alpha_1 t}$$

As $L_0\alpha_1 t$ is very small compared to L_0 the effect on the equation of neglecting it is insignificant. Thus

$$\text{strain} = \frac{x_1}{L_0} = \frac{\sigma_1}{E_1}$$

and so

$$x_1 = \frac{L_0\sigma_1}{E_1}$$

Similarly for material 2

$$x_2 = \frac{L_0\sigma_2}{E_2}$$

Substituting these back into Equation (2.3) gives

$$L_0 a_2 t - L_0 a_1 t = \frac{L_0\sigma_2}{E_2} + \frac{L_0\sigma_1}{E_1}$$

This equation can be simplified to

$$(\alpha_2 - \alpha_1)t = \frac{\sigma_2}{E_2} + \frac{\sigma_1}{E_1} \tag{2.4}$$

In addition to the above relationship, another equation can be developed linking the stresses in the two materials.

tensile force on 1 = compressive force on 2

But

tensile force on 1 = tensile stress σ_1 × area A_1

and

compressive force on 2 = compressive stress σ_2 × area A_2

The areas are the cross-sectional areas of the materials. Hence

$$\sigma_1 A_1 = \sigma_2 A_2 \tag{2.5}$$

The stresses acting on the two materials can be obtained by solving the two simultaneous equations (2.4) and (2.5).

The above analysis gives the stresses acting on the materials due solely to a temperature change. If the materials are also loaded then the total stress on a material is due to the combined effects of the two stresses; they can be just added with due regard being paid to whether they are compressive or tensile. A tensile stress is normally regarded as being positive and a compressive stress as negative. Thus a tensile stress of 0.10 kN/mm^2 due to loading when added to a compressive stress of 0.05 kN/m^2 due to temperature gives a total stress of +0.10 – 0.05 = +0.05 kN/mm^2, a tensile stress.

Example 22 A steel tube with an external diameter of 35 mm and an internal diameter of 30 mm has a brass rod of diameter 20 mm inside it and rigidly joined to it at each end. At 15 °C, when the materials were joined, there were no stresses in the materials. What will be the longitudinal stresses produced when the temperature is raised to 100 °C? The steel has a modulus of elasticity of 210 kN/mm^2 and a coefficient of linear expansion of 11 × 10^{-6}/°C. The brass has a modulus of elasticity of 120 kN/mm^2 and a coefficient of linear expansion of 18 × 10^{-6}/°C.

$$(\alpha_2 - \alpha_1)t = \frac{\sigma_2}{E_2} + \frac{\sigma_1}{E_1}$$

Taking the steel to be material 1, gives

$$(18 - 11)10^{-6} \times (100 - 15) = \frac{\sigma_2}{120} + \frac{\sigma_1}{210}$$

The unit of the stresses will be kN/mm^2.

$$595 \times 10^{-6} = \frac{\sigma_2}{120} + \frac{\sigma_1}{210}$$

$$\sigma_1 A_1 = \sigma_2 A_2$$

Hence

$$\sigma_1 \times \frac{\pi}{4}(35^2 - 30^2) = \sigma_2 \frac{\pi}{4} 20^2$$

$$325\sigma_1 = 400\sigma_2$$

The two simultaneous equations can now be solved.

$$595 \times 10^{-6} = \frac{325\sigma_1}{400 \times 120} + \frac{\sigma_1}{210}$$

$$400 \times 120 \times 210 \times 595 \times 10^{-6} = 325 \times 210\sigma_1 + 400 \times 120\sigma_1$$

$\therefore \quad \sigma_1 = 0.052$ kN/mm^2 to two significant figures.

$\sigma_2 = \frac{325}{400}\sigma_1 = 0.042$ kN/mm^2 to two significant figures

The material with the smaller coefficient of linear expansion is in tension and the material with the larger coefficient is in compression. Thus the stress in the steel σ_1 is tensile and the stress in the brass σ_2 is compressive.

Example 23 What will be the stresses in the steel tube and the brass rod in the above example if an axial compressive load of 80 kN is applied?

Ignoring the temperature change for the moment and just considering the effect of the load,

Total force = stress on 1 × area of 1 + stress on 2 × area of 2

$$80 = \sigma_1 \times \frac{\pi}{4}(35^2 - 30^2) + \sigma_2 \times \frac{\pi}{4}20^2$$

$$320 = \sigma_1 \times \pi \times 325 + \sigma_2 \times \pi \times 400$$

Because the forces are kN and the units of area are mm², the unit of stress will be kN/mm²

$$\frac{\text{stress on 1}}{E \text{ of } 1} = \frac{\text{stress on 2}}{E \text{ of } 2}$$

$$\frac{\sigma_1}{210} = \frac{\sigma_2}{120}$$

The two simultaneous equations can be solved to give the two stresses.

$$320 = \frac{210}{120}\sigma_2 \times \pi \times 325 + \sigma_2 \times \pi \times 400$$

$\sigma_2 = 0.11$ kN/mm² to two significant figures.

$\sigma_1 = \frac{210}{120}\sigma_2 = 0.18$ kN/mm² to two significant figures.

Both these stresses are compressive. Thus the stress on the steel due to both the load and the temperature change is

$0.052 - 0.11 = -0.058$ kN/mm².

The steel is thus in compression. The stress on the brass due to both the load and the temperature change is

$-0.042 - 0.18 = -0.22$ kN/mm² to two significant figures.

The brass is thus in compression.

REINFORCED AND PRESTRESSED CONCRETE

The tensile strength of concrete is generally about 5 N/mm², considerably lower than metals. The compressive strength, i.e. the stress needed to fracture the concrete when in compression, is generally about 65 N/mm². Concrete is thus much stronger in compression than in tension. To overcome the low tensile strength of concrete, it is reinforced with steel rods, the tensile stress then being shared between the steel and the concrete. This has the effect of reducing the tensile stress applied to the concrete, so enabling the reinforced concrete to carry a higher tensile load than the concrete alone could carry.

For the composite of a steel rod in a column of concrete then (see page 23):

$$\frac{\text{stress on concrete}}{E \text{ of concrete}} = \frac{\text{stress on steel}}{E \text{ of steel}}$$

Because the elastic modulus of steel is higher than that of concrete, the steel takes a bigger share of the stress than the concrete. Typical values of the elastic modulus are 200 kN/mm^2 for the steel and 20 kN/mm^2 for the concrete. The elastic modulus of the steel is about ten times higher than that of the concrete. Thus

stress on steel = 10 × stress on concrete

The total force that can be carried by the composite is given by (see page 23):

Total force = stress on concrete × its area + stress on steel × its area

In the absence of the steel the total force possible is limited by the stress at which the concrete fails. Thus if the concrete was used in tension then the tensile strength of about 5 N/mm^2 would indicate that the maximum force possible would be

Total force = 5 × area of concrete N/mm^2 × area units

The addition of the term to the equation for the steel enables this total force to be considerably increased.

Total force = 5 × area of concrete + stress on steel × its area

As the stress on the steel will be ten times the stress on the concrete, because of the higher modulus of elasticity of the steel, then the maximum force possible before the concrete fails would be

Total force = 5 × area of concrete + 50 × area of steel

The addition of quite a small area of steel to the concrete can thus raise the force that can be carried by the concrete by a considerable amount.

In reinforced concrete, the steel rods are embedded in the concrete, the concrete being poured round the rods and allowed to set. The rods are placed so that they are in the part of the composite most likely to undergo tension, the concrete being much weaker in tension than in compression (Figure 2.7(a)).

Another form of reinforced concrete is *prestressed concrete*. In this composite, the concrete is put into a state of compression by means of steel wires (Figure 2.7(b)). In one form of prestressed concrete, the

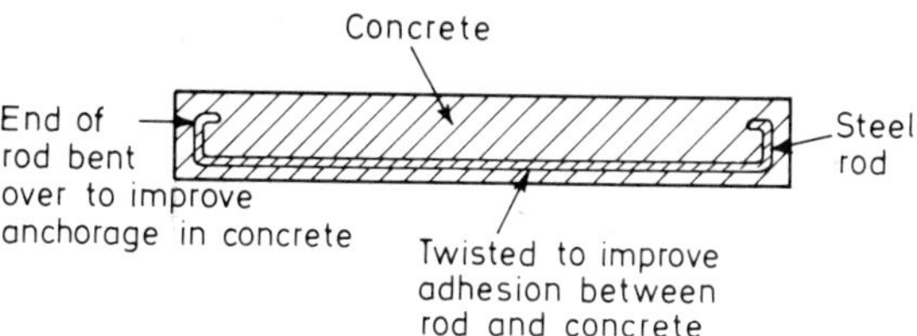

(a) A reinforced concrete beam

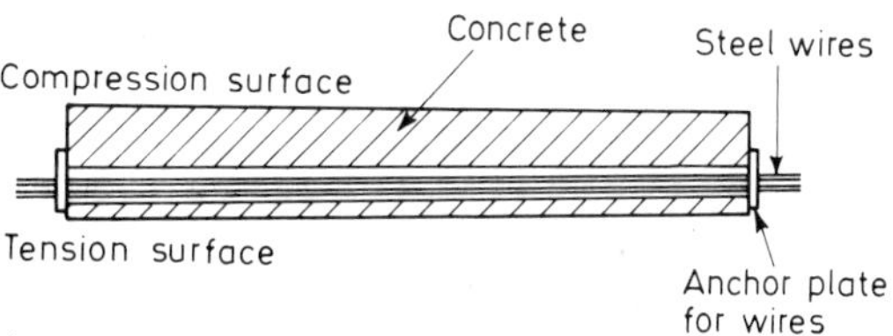

(b) A prestressed concrete beam

Figure 2.7

steel wires are put in tension between two clamps before being surrounded by the wet concrete. When the concrete has set the clamps are removed and the wires then apply stresses to the concrete which result in the concrete being put into compression. Another method of putting the concrete into compression is post-tensioning of the wires. In this method the concrete sets round tubes through which wires are inserted when the concrete is set. These wires are then stretched and anchored, so putting the concrete in compression. When prestressed concrete is in service, the initial compressive stresses in the concrete have to be overcome before the concrete is put into tension.

FIBRE REINFORCED MATERIALS

Reinforced concrete is just one example of the way in which the properties of a material, i.e. concrete, can be changed by the incorporation of another material. The incorporation of fibres in materials can lead to very marked changes in the properties of the material. Glass fibre reinforced plastics have been used for a number of years and now carbon reinforced plastics and metals are being used with the likelihood of many more reinforced materials being introduced very soon. Fibre reinforced materials have been termed 'the materials of the future'.

Aluminium has a modulus of elasticity of about 70 kN/mm^2. Carbon fibres can have a modulus as high as 400 kN/mm^2. Combine the two into a composite (Figure 2.8) and it is possible to have a carbon fibre reinforced aluminium with a modulus of elasticity of about 300 kN/mm^2. This composite is much stiffer than the aluminium alone. The composite also has a much higher tensile strength.

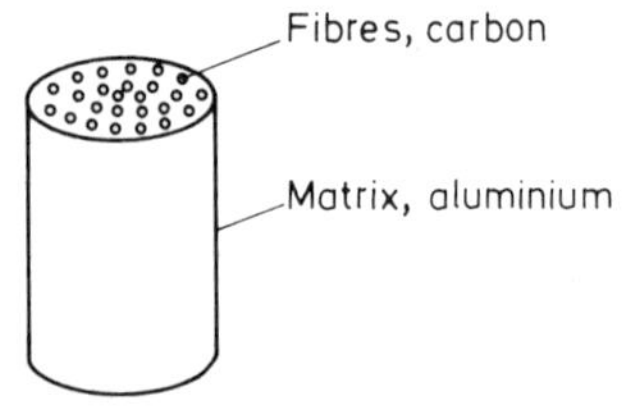

Figure 2.8 Carbon fibre reinforced aluminium

In the same way as in the consideration of the composite on page 23, we have for a composite consisting of fibres in a supporting base or matrix, with the fibres running parallel to the axis along which the load is applied:

$$\text{Total force on composite} = \text{stress on fibre} \times \text{area of fibre} + \text{stress on matrix} \times \text{area of matrix}$$

Dividing throughout by the total cross-sectional area of the composite gives

$$\frac{\text{Total force on composite}}{\text{area of composite}} = \frac{\text{stress on fibre} \times \text{area of fibre}}{\text{area of composite}} + \frac{\text{stress on matrix} \times \text{area of matrix}}{\text{area of composite}}$$

But the total force on the composite divided by the area of the composite is the stress acting on the composite. The area of the fibre divided by the area of the composite is the fraction of the cross-sectional area that is fibre. The area of the matrix divided by the area of the composite is the fraction of the cross-sectional area that is matrix. Thus the equation can be written as

$$\text{Stress on composite} = \text{stress on fibre} \times \text{area fraction fibre} + \text{stress on matrix} \times \text{area fraction matrix.}$$

The strain experienced by the composite is the same as that experienced by both the fibres and the matrix. Dividing throughout by the strain changes the equation into one between the modulii of elasticity. Thus

$$E \text{ of composite} = E \text{ of fibre} \times \text{area fraction fibre} + E \text{ of matrix} \times \text{area fraction matrix.}$$

Example 24 Carbon fibres with a modulus of elasticity of 400 kN/mm^2 are used to reinforce aluminium with a modulus of 70 kN/mm^2. What will be the resulting modulus of elasticity of the composite if half of the composite is taken up by the fibre?

$$E \text{ of composite } = E \text{ of fibre} \times \text{area fraction fibre} + E \text{ of matrix} \times \text{area fraction matrix}$$

Thus

$$E \text{ of composite } = 400 \times \tfrac{1}{2} + 70 \times \tfrac{1}{2} \quad \text{kN/mm}^2$$
$$= 235 \text{ kN/mm}^2$$

POISSON'S RATIO

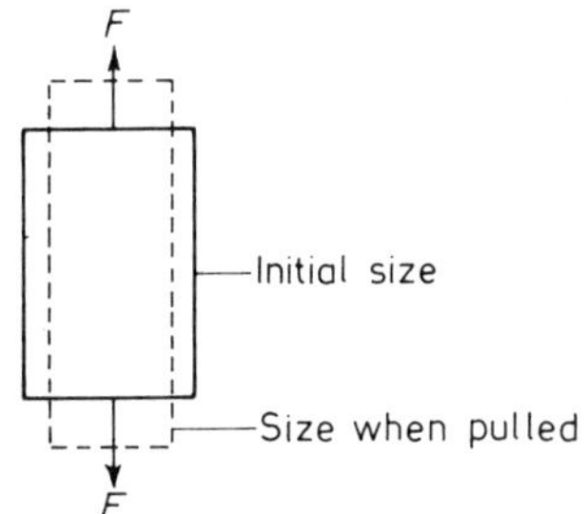

Figure 2.9 A longitudinal strain results in a transverse strain

When a piece of metal is pulled, i.e. put into tension, then there is a transverse contraction of the material (Figure 2.9). When stretching a rubber band the same thing happens, the band gets noticeably thinner, i.e. a longitudinal strain has given rise to a transverse strain. The ratio of the transverse strain to the longitudinal strain is called *Poisson's ratio, ν.*

$$\text{Poisson's ratio} = \frac{\text{transverse strain}}{\text{longitudinal strain}}$$

As it is a ratio, ν has no units. For most engineering materials Poisson's ratio is about 0.3. The following are some typical values.

Material	*Poisson's ratio*
Mild steel	0.31
Cast iron	0.25
Aluminium alloys	0.30
Copper	0.35
Brass	0.35

Because the transverse strain is compressive when the longitudinal strain is tensile, and tensile when the longitudinal strain is compressive, the equation for Poisson's ratio is sometimes written with a minus sign.

$$\text{Poisson's ratio} = -\frac{\text{transverse strain}}{\text{longitudinal strain}}$$

This means that when one of the strains is positive, i.e. tensile, the other must be negative, i.e. compressive.

Example 25 A bar of mild steel of length 100 mm is extended by 0.01 mm, by how much will the width of the bar contract if initially the bar had a width of 10 mm? Poisson's ratio can be taken as 0.31.

$$\text{Longitudinal strain} = \frac{0.01}{100} = 0.0001$$

$$\text{Poisson's ratio} = -\frac{\text{transverse strain}}{\text{longitudinal strain}}$$

$$0.31 = -\frac{\text{transverse strain}}{0.0001}$$

Hence

$$\text{transverse strain} = -0.000\,031$$

But

$$\text{transverse strain} = \frac{\text{change in width}}{\text{width}}$$

Hence

$$\text{change in width} = -0.000\,031 \times 10$$
$$= -0.000\,31 \text{ mm}$$

The minus sign indicates that the width is reduced by the longitudinal strain.

SHEAR STRESS AND STRAIN

Figure 2.4 shows how forces can produce tensile and compressive stresses, the effect of these stresses being to cause the material to extend or contract. There is another way in which materials can be deformed. When a material has forces applied in such a way that one layer of the material tends to slide over an adjacent layer, then the forces are said to be shearing and be *shear* forces. Figure 2.10 illustrates this action. Compare it with Figure 2.4.

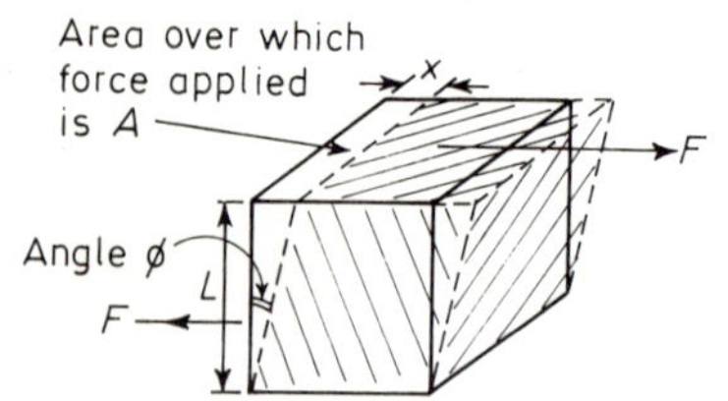

Figure 2.10 A cube subject to shear forces

With tension and compression, the force is at right angles to the area upon which it is acting. Tensile or compressive stress is the force divided by the area at right angles to the line of action of the force. With shear, the force is in the same plane as the area to which it is applied. The effect of the force depends on the area over which it is applied, just as with tension and compression, and for the same force per unit area the same deformation is produced in a material if the length L is the same (see Figure 2.10). Thus if the area A is doubled then for the same deformation the force has to be doubled. The force per unit area is called the *shear stress,* τ.

$$\text{Shear stress} = \frac{\text{force}}{\text{area}}$$

The unit of shear stress is N/m^2 or Pa.

With tensile and compressive stresses the deformations produced are changes in length, extensions and compressions. With shear stress the deformation produced is an angular deformation ϕ. For the same shear stress with a material the same angular deformation is produced regardless of L. *Shear strain* is defined as the angle ϕ.

$$\text{Shear strain} = \phi$$

As ϕ is in radians, shear strain has no units.

There is an alternative manner of considering shear strain which is in appearance more comparable with the definitions of tensile and compressive strain.

$$\text{Shear strain} = x/L$$

If x and L are in the same length units, e.g. m, then shear strain has no unit as it is a ratio.

The above two statements of shear strain are the same if the second statement is restricted to x being considerably smaller than L, usually the case in practice.

$$\frac{x}{L} = \tan \phi$$

But for small angles $\tan \phi$ is virtually the same as ϕ expressed in radians.

ϕ (degrees)	$\tan \phi$	ϕ (radians)
1	0.0175	0.0175
2	0.0349	0.0349
3	0.0524	0.0524
4	0.0699	0.0698

For most metals the shear strain is ϕ less than 1° when the deformation is elastic and not permanent. The symbol for shear is γ.

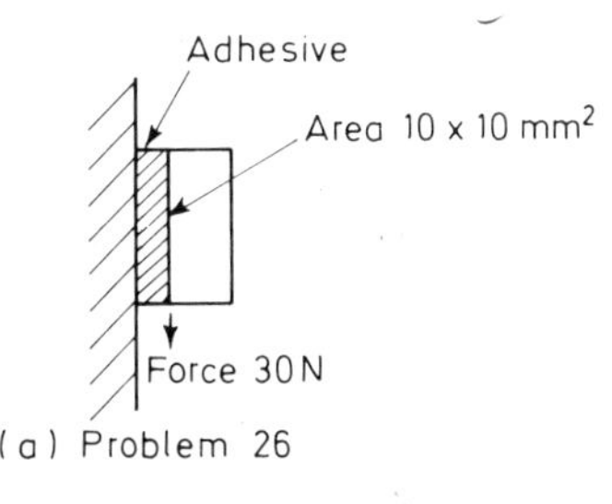

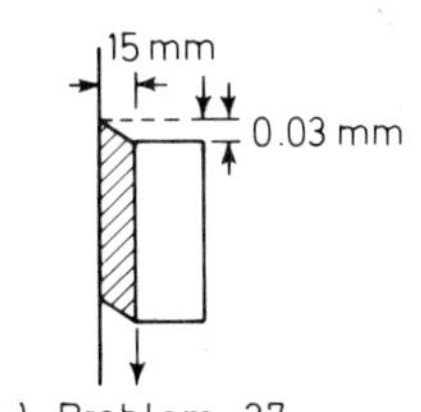

Figure 2.11

Example 26 Figure 2.11 shows a component which is attached to a vertical surface by means of adhesive. The area of the adhesive in contact with the component is 100 mm^2, being a square 10 mm × 10 mm. The weight of the component results in a force of 30 N being applied to the adhesive-component interface. What is the shear stress?

$$\text{Shear stress} = \frac{\text{force}}{\text{area}}$$

$$= \frac{30}{100} \quad \frac{\text{N}}{\text{mm}^2}$$

$$= 0.3\ \text{N/mm}^2$$

Example 27 If the component in Figure 2.11 sags by 0.03 mm under the action of the shear stress at the adhesive–component interface, the adhesive having a thickness of 15 mm, what is the shear strain?

$$\text{Shear strain} = \frac{0.03}{15} \quad \frac{\text{mm}}{\text{mm}}$$

$$= 0.002$$

MODULUS OF RIGIDITY

There are many materials for which the direct strain is proportional to the direct stress within some limiting stress. Such materials are said to obey Hooke's law. Such materials also have a shear strain proportional to the shear stress within some limiting shear stress (Figure 2.12). For this region

$$\frac{\text{shear stress}}{\text{shear strain}} = \text{a constant, } G$$

Figure 2.12 Shear stress/shear strain graph

The constant G is called the *modulus of rigidity* or the *shear modulus*. It has units of N/m^2 if the stress is in units of N/m^2. The following are some typical values of the modulus.

Material	*Modulus of rigidity*/kN mm^{-2}
Mild steel	76
Cast iron	33
Aluminium alloys	27
Brass	37
Copper	46
Rubber	0.001

The modulus of elasticity is a measure of the stiffness of a material, the greater the modulus of elasticity the more difficult it is to bend the material. The modulus of rigidity is a measure of the rigidity or ease of twisting of a material, the greater the modulus of rigidity the more difficult it is to twist the material (Figure 2.13). Twisting a rod of a material involves shear forces. The twisting of rods and tubes is considered in more detail in Chapter 4.

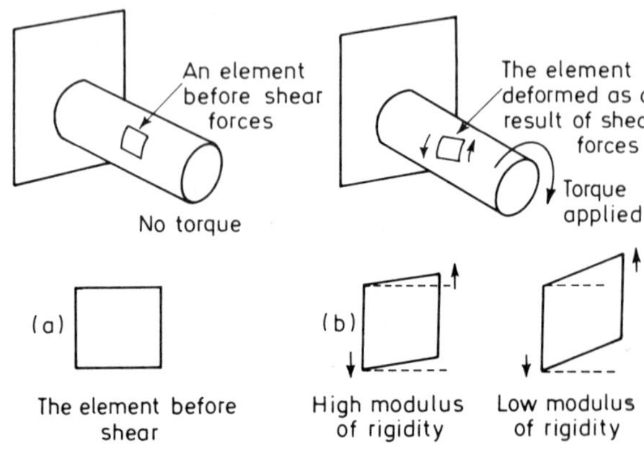

Figure 2.13 (a) The effect of twisting a shaft is to apply shear forces to each element. (b) The effect on an element of the same shear forces when the materials are of different modulii of rigidity. The lower the modulus, the easier it is to twist

SHEAR STRENGTH

The *shear strength* of a material is the maximum shear stress that material can withstand before failure by shear occurs (Figure 2.9). The following are some typical values.

Material	*Shear strength*/N mm^{-2}
Mild steel	200
Aluminium alloy	90
Brass	250

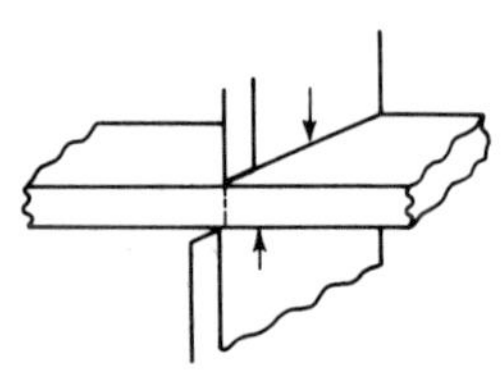

Figure 2.14 A guillotine cropping plate

Every time a guillotine is used to crop a material, perhaps plate in a steel rolling mill, shear stresses are being applied at a value equal to the maximum shear stress for that material (Figure 2.14). The area over which the shear forces are being applied is the cross-sectional area of the plate being cropped. The greater the cross-sectional area of the material the greater the force that has to be applied by the guillotine in order for the ultimate shear stress to be realized.

When a punch is used to punch holes in a material (Figure 2.15) then again shear stresses are being applied to a value equal to the maximum shear stress for that material. The area over which the punch force is applied is the plate thickness multiplied by the perimeter of the hole being punched.

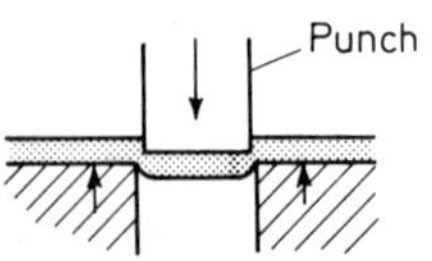

Figure 2.15 A punch in action

Example 28 A plate of mild steel 1000 mm wide and 0.8 mm thick is to be cropped by a guillotine. What is the force which the guillotine has to apply if the shear strength of the steel is 200 N/mm²?

$$\text{Shear strength} = \frac{\text{force necessary}}{\text{cross-sectional area}}$$

Hence

$$\text{force necessary} = 200 \times 1000 \times 0.8 \quad \text{N/mm}^2 \times \text{mm}^2$$

$$\text{force necessary} = 160 \text{ kN}$$

Example 29 What is the maximum diameter hole that can be punched in an aluminium plate of thickness 14 mm if the punching force is limited to 50 kN? The shear strength of the aluminium can be taken as 90 N/mm²

$$\text{shear strength} = \frac{\text{punch force necessary}}{\text{area being sheared}}$$

The area being sheared is the perimeter of the hole, i.e., πd, multiplied by the plate thickness. Hence

$$\pi d \times 14 = \frac{50 \times 1000}{90}$$

d = 130 mm to two significant figures

PROBLEMS

1. A tensile force of 80 kN is applied to a tensile test specimen of cross-sectional area 50 mm². What is the tensile stress acting on the specimen in units of (a) kN/mm² (b) N/m² (c) GPa?

2. A pipe has an external diameter of 35 mm and an internal diameter of 30 mm. What is the tensile stress acting on the pipe if the pipe is subject to a tensile load of 80 kN? Give the answer in units of (a) kN/mm² (b) N/m².

3. A connecting link has a cross-sectional area of 300 mm² and is subject to a tensile force of 30 kN. What is the tensile stress produced in kN/mm²?

4. A tensile force acting on a specimen of length 300 mm causes it to extend by 2 mm in the direction of the force. What is the strain produced?

5. A 0.1% strain is produced by the action of a force on a bar of steel. What is the extension if the bar has a length of 500 mm?

6. The modulus of elasticity of mild steel is 210 kN/mm². What is the value of this modulus in units of (a) N/m² (b) kN/m² (c) Pa (d) GPa?

7. A tensile test specimen has a gauge length of 100 mm and a cross-section of diameter 11.28 mm. If the material is an aluminium alloy of modulus of elasticity of 70 kN/mm² what extension of the gauge length might be expected if the tensile load applied to the specimen was 20 kN?

8. Calculate the strain that will be produced in a bar of copper when subject to a tensile stress of 2 N/mm^2. The copper has a modulus of elasticity of 120 kN/mm^2.

9. A tensile test specimen has a diameter of 7.98 mm and a gauge length of 50 mm. The following are results obtained during the tensile test. Estimate the value of the modulus of elasticity of the material.

Load/kN	0	20	40	60	80
Extension/mm	0	0.0024	0.0048	0.0073	0.0096

10. A rubber band has a rectangular cross-section of 5 mm × 1 mm and a length 80 mm (this is a rubber band used for holding papers together and so is a closed loop. The actual length of rubber round the entire loop is thus 2 × 80 mm). When I pull the band between my hands I stretched it by 100 mm. If Hooke's law is obeyed and if the modulus of elasticity is 7 N/mm^2, what is the force I applied?

11. Deduce the modulus of elasticity and the 0.1% proof stress for the material which gave the following tensile test results. The diameter of the test piece was 5.64 mm and the gauge length 25 mm.

Force/kN	0	0.5	1.0	1.5	2.0	2.5	3.0	3.5	4.0	4.5 (broke)
Extension/mm	0	0.0063	0.0125	0.0188	0.0250	0.0340	0.0431	0.0573	0.0751	0.0974

12. Deduce the modulus of elasticity and the 0.1% proof stress for the material which gave the following tensile test results. The diameter of the test piece was 11.28 mm and the gauge length 100 mm.

Force/kN	0	5	10	15	20	25	30	35	40	40.3 (broke)
Extension/10^{-3} mm	0	1.61	3.30	5.02	6.73	8.50	10.30	12.59	26.41	33.90

13. Deduce the modulus of elasticity and the yield stress for the material which gave the following tensile test results. The diameter of the test piece was 15.96 mm and the gauge length was 200 mm.

Force/kN	0	20	40	60	80	100	120	140 (broke)
Extension/mm	0	0.064	0.128	0.195	0.260	0.325	0.387	0.890

14. A material has a yield stress of 200 N/mm^2. At what force should a bar of the material yield if it has a cross-section of 30 mm × 20 mm?

15. A mild steel strut has a cross-section of 200 mm × 100 mm. What should be the maximum permissible force if the steel has a yield stress of 200 N/mm^2 and a factor of safety of 4 is used?

16. Calculate the tensile strengths of the materials giving the tensile test data in questions 11, 12 and 13.

17. An aluminium alloy has a tensile strength of 200 N/mm^2. What force is needed to break a bar of the alloy with a cross-sectional area of 250 mm^2?

18. The tensile strength of the rubber used for a rubber band is 30 N/mm^2. Estimate the force needed to break the rubber band referred to in question 10.

19. A strut with a cross-sectional area of 400 mm^2 has a compressive load of 50 kN applied to it. Calculate the compressive stress acting on the strut.

20. Calculate the minimum diameter of a circular bar of mild steel if it is to withstand a compressive stress of 30 kN. The yield stress of the mild steel is 200 N/mm^2 and a factor of safety of 4 is to be used.

21. Determine the thickness and diameter of a rubber pad of circular cross-section that can be used to support a machine if it is to carry a load of 4 kN and to compress by no more than 4 mm under this load. The stress in the rubber is not to exceed 250 kN/m^2, this being the safe working stress for the material. Young's modulus for the rubber is 2 N/mm^2.

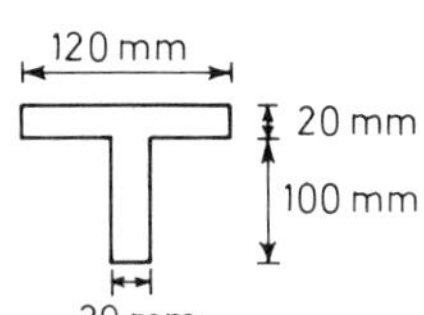

Figure 2.16

22. Figure 2.16 shows the cross-section of a T-section made of steel. What compressive load can be applied to a girder having such a section if the load is applied at right angles to the area shown and the maximum permissible stress is 50 N/mm^2?

23. Calculate the amount a pipe will shorten when subject to an axial compressive load of 20 kN. The pipe has an external diameter of 70 mm and an internal diameter of 60 mm. The length of the pipe is 3.0 m and the pipe material has a modulus of elasticity of 200 N/mm^2.

24. What forces are required to stop a copper rod 1.5 m long and 30 mm diameter expanding when the temperature rises by 20 °C. The coefficient of linear expansion of the copper is 17×10^{-6}/°C and the modulus of elasticity 120 N/mm^2.

25. An aluminium alloy bar is constrained between two rigid supports and prevented from contracting or expanding. Calculate the stress developed in the bar as a result of a drop in temperature of 15 °C. The coefficient of linear expansion can be taken as 23×10^{-6}/°C and the modulus of elasticity 70 kN/mm^2.

26. A composite bar is made of a copper tube containing an inner close fitting steel core. The metals are bonded together so that relative movement cannot occur. The steel core has a diameter of 35 mm, the copper tube a diameter of 45 mm. Calculate the stress in each material when an axial load of 90 kN is applied to the composite bar. The steel has a modulus of elasticity of 210 kN/mm^2 and the copper a modulus of 120 kN/mm^2.

27. A pile used for the foundation of a building consists of a cast iron cylinder filled with concrete. The cylinder has a diameter of 250 mm and a thickness of 25 mm. The pile is to carry a load of 100 kN. What will be the stresses acting on the concrete and the cast iron? The modulus of elasticity of the cast iron is 100 kN/mm^2 and that of the concrete 20 kN/mm^2.

28. A reinforced concrete column has a circular cross-section of diameter 300 mm and a height of 5 m. If the column is to carry an axial load of 500 kN without the stress in the concrete exceeding 4 MN/m^2, what must be the minimum diameter of a steel reinforcement rod? If steel reinforcement rods are only available with a diameter of 25 mm, how many will be needed? The modulus of elasticity of the steel can be taken as 210 kN/mm^2 and that of the concrete as 20 kN/mm^2.

29. A bar of mild steel is used to reinforce a timber tie. The timber has a rectangular cross-section of 300 mm by 130 mm and the steel bar a rectangular cross-section of 40 mm by 10 mm. What are the stresses in the timber and the steel when the tie carries an axial load of 1.2 MN?

The modulus of elasticity of the timber is 15 kN/mm^2 and that of the steel 210 kN/mm^2.

30. A cast iron pipe of external diameter 65 mm and internal diameter 45 mm has a steel bolt of diameter 35 mm passing through it. The bolt is kept within the pipe by means of nuts at each end. By how much further than just in contact with the end should a nut be tightened in order to stress the pipe to 45 MN/m^2? The modulus of elasticity for the cast iron can be taken as 100 kN/mm^2 and that for the steel 210 kN/mm^2.

31. A steel tube with an external diameter of 40 mm and an internal diameter of 35 mm has a brass rod of diameter 25 mm inside it and rigidly joined to it at each end. The materials were joined together at a temperature of 250 °C and can be assumed to have no longitudinal stresses present at that temperature. What will be the longitudinal stresses produced in the materials when they cool to 20 °C? The steel has a modulus of elasticity of 210 kN/mm^2 and a coefficient of linear expansion of 11×10^{-6}/°C. The brass has a modulus of elasticity of 120 kN/mm^2 and a coefficient of linear expansion of 18×10^{-6}/°C.

32. A steel plate is clad on both sides by sheets of copper. The steel plate is 10 mm thick and the copper cladding 2 mm thick. Calculate the change in stress in both materials for each Celsius degree rise in temperature. The steel has a modulus of elasticity of 210 kN/mm^2 and a coefficient of linear expansion of 11×10^{-6}/°C and the copper has a modulus of elasticity of 120 kN/mm^2 and a coefficient of linear expansion of 17×10^{-6}/°C.

33. What will be the effect on the answer to question 26 if the temperature increases by 15 °C? The coefficient of linear expansion of the copper is 17×10^{-6}/°C and that of the steel 11×10^{-6}/°C.

34. The following are the coefficients of expansion of some common building materials.

Material	*Linear coefficient of expansion*/10^{-6} °C^{-1}
Brick	3 to 9
Cement and concrete	10 to 14
Steel	11
Wood, along the grain	3 to 5
Wood, across the grain	35 to 60

(a) Reinforced concrete consists of steel rods in concrete. What might be the effect of temperature changes on the stresses in such a composite?
(b) A house as well as using wooden beams may also have steel girders. What might be the effect of a composite arrangement where wooden beams and steel girders are linked together?

35. A bar of aluminium alloy of length 2 m is extended by 0.5 mm. By how much will the width of the bar change if it has an initial width of 100 mm? Poisson's ratio for the alloy is 0.30.

36. A copper rod of diameter 50 mm and length 500 mm is subject to a compressive axial load of 10 kN. By how much will the rod be compressed longitudinally and by how much transversally by the load?

The modulus of elasticity is 120 kN/mm^2 and Poisson's ratio for copper is 0.35.

37. A connecting link has a length of 500 mm and a cross-section of 30 mm by 10 mm and is subject to a tensile axial force of 35 kN. What are the dimensions of the link when under load? The modulus of elasticity of the material used is 210 kN/mm^2.

38. A cantilever projects 80 mm from its clamped point. The cantilever has a cross-section of 40 mm by 20 mm. If a force of 50 kN is applied at the free end of the cantilever, in the plane of the end face, what is the shear stress?

39. Calculate the shear strain that results from the application of the force in problem 38 if the modulus of rigidity of the material is 76 kN/mm^2. What will be the deflection of the end of the cantilever?

40. Calculate the maximum thickness of plate that can have a hole punched in it by a punch of diameter 20 mm if the shear strength of the plate is 200 N/mm^2 and the maximum force that can be exerted by the punch is 100 kN.

41. Calculate the force needed to shear a mild steel plate of thickness 0.7 mm and width 1 m if the shear strength of the steel is 200 N/mm^2.

42. Calculate the force needed to crop bars of circular cross-section with a diameter of 30 mm if the shear strength is 90 N/mm^2.

3 Bending beams

BEAMS

Beams are in very common use. You may probably at this present moment be supported by a beam if you are not in a ground floor room. Over your head, if you are indoors, there are probably beams supporting the floor above or the roof. The beam is an important device in structural engineering and beam theory might well be considered the backbone of structural engineering.

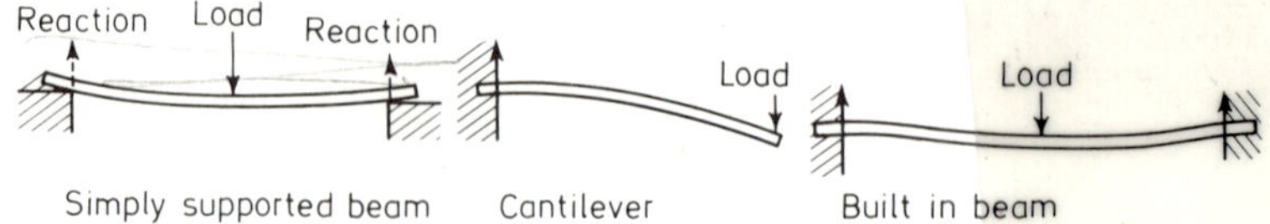

Figure 3.1 Simple types of beams

A *beam* can be defined as a structural member that offers resistance to bending caused by applied loads. Figure 3.1 shows some simple types of beams. Beams should not be just regarded as rectangular cross-section bars of wood or metal; A beam can be of any cross-sectional shape. A very common form is the I beam (Figure 3.2).

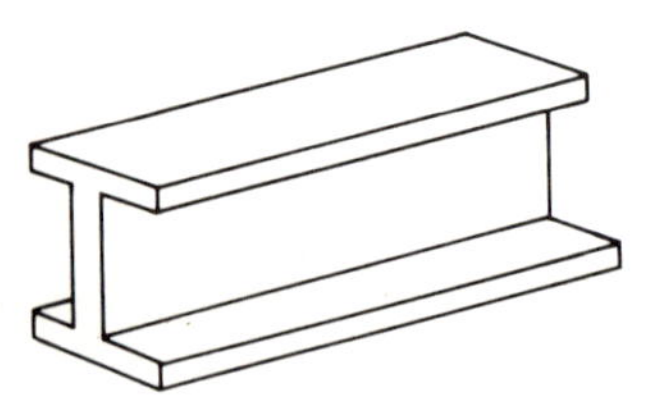

Figure 3.2 I beam

Beams deflect under load and become curved. Thus for the beam shown in Figure 3.3, the upper surface of the beam becomes smaller than its initial length and the lower surface becomes extended. The upper surface is thus in compression and the lower surface in tension. At some intermediate position between the upper and lower surfaces there must therefore be a layer which is the same length in the bent beam as in the unbent beam. This layer, or surface, is called the *neutral surface* and the line in which the neutral surface cuts the cross-section of the beam is called the *neutral axis*. Because this neutral surface is the same length under load as with no load it is under no strain and hence no stress. Thus while the upper beam surface is under compressive stress and the lower beam surface tensile stress, the neutral surface is unstressed.

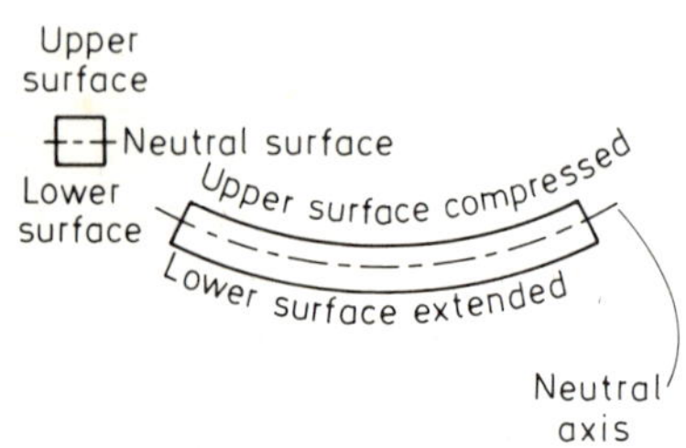

Figure 3.3 Beam curving under a load

The parts of the beam that are the most stressed are those the furthest from the neutral surface, in the case of Figure 3.3 the upper and lower surfaces of the beam. Because these surfaces are the highest stress regions it is necessary to have as much material in these regions as possible if the beam is to bend as little as possible under the action of a load. There is no great need for material near the neutral surface as there is very little stress there. Hence the widespread use of the I beam in structural engineering, a beam with most of the material in the high stress region and least material in the low stress region. There is no point in paying for material that is not necessary.

BENDING MOMENT

One of the conditions for static equilibrium of an object is that the sum of the anticlockwise moments equals the sum of the clockwise moments (see Chapter 1). Thus if we consider a simply supported beam, as in Figure 3.4, if the beam is in equilibrium the anticlockwise

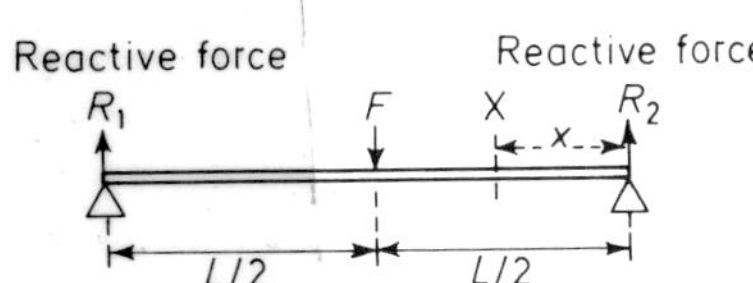

Figure 3.4

moment about any axis must equal the clockwise moment about the same axis.

Taking moments about the axis at which reactive force R_1 operates gives:

$$\text{Anticlockwise moments} = R_2 \times L$$

$$\text{Clockwise moments} = F \times \frac{L}{2}$$

Hence

$$R_2 \times L = F \times \frac{L}{2}$$

$$R_2 = \frac{F}{2}$$

Taking moments about the axis at which reactive forces R_2 operates gives:

$$\text{Anticlockwise moments} = F \times \frac{L}{2}$$

$$\text{Clockwise moments} = R_1 \times L$$

Hence

$$R_1 \times L = F \times \frac{L}{2}$$

$$R_1 = \frac{F}{2}$$

Both the reactive forces are thus $F/2$.

This checks with the other equilibrium condition that the upwards directed forces equal the downwards directed forces.

$$F = R_1 + R_2$$

$$F = \frac{F}{2} + \frac{F}{2}$$

At a particular section X of the beam there will be an anticlockwise moment due to the reactive force R_2 (Figure 3.5(a)). The moment is R_2x or $Fx/2$. This moment is endeavouring to rotate the beam to the right about the section X.

At the same section X of the beam there will also be a moment due to the forces to the left of X (Figure 3.5(a,b,c)). The anticlockwise moment due to the force F is $F(L/2 - x)$. The clockwise moment due to the reactive force R_1 is $\frac{1}{2}F(L - x)$. There is a net clockwise moment of

$$\frac{F}{2}(L - x) - F\left(\frac{L}{2} - x\right) = \frac{Fx}{2}$$

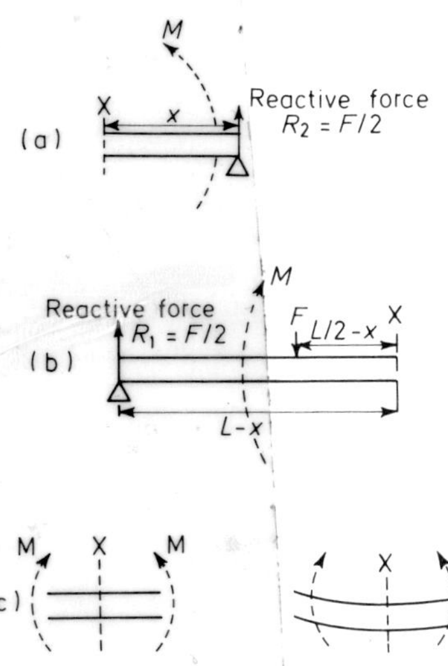

Figure 3.5 (a) Moment due to force to the right of X. (b) Moment due to forces to the left of X. (c) Moments causing the bending at X

The moment about X due to the force to the right of X (Figure 3.5(a)) and the moment about X due to the forces to the left of X (Figure 3.5(b)) are both tending to cause the beam at X to bend (Figure 3.5(c)). Both the moment for the right side of X and the left side of X are $Fx/2$. This moment is called the bending moment about that section.

The *bending moment* at a transverse section of a beam is the algebraic sum of all the moments about the section of all the forces acting on one (either) side of the section concerned.

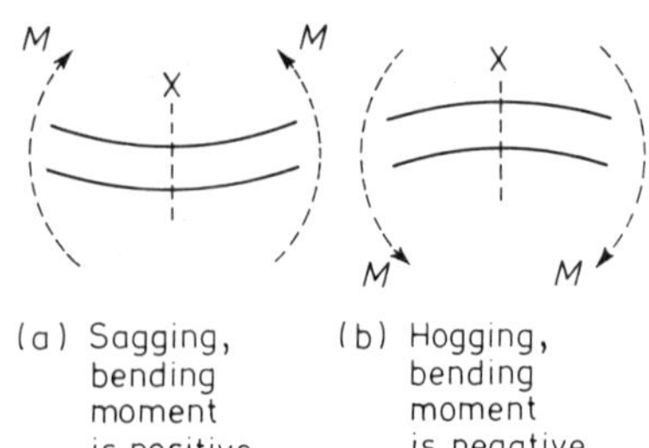

Figure 3.6

A bending moment is considered to be positive when the moment on the left of the section concerned is clockwise and that on the right anti-clockwise. Such a bending moment produces sagging of the beam (Figure 3.6(a)). A bending moment is considered to be negative when the moment on the left of the section concerned is anticlockwise and that on the right clockwise. Such a bending moment produces hogging of the beam (Figure 3.6(b)).

Example 1 A beam 4 m long rests on supports at each end. A load of 500 N is applied to the beam centre. What will be the bending moment a distance of (a) 1 m and (b) 2½ m from the right hand end. Figure 3.4 represents the beam.

Because the beam is centrally loaded the reactions at each end will be the same, the sum of the two reactions being equal to the applied load. For equilibrium the upwards directed forces must equal the downwards directed forces. Hence the reactions at each end are 250 N.

(a) The bending moment due to the part of the beam to the right of the point concerned will be

$$\text{Bending moment} = +250 \times 1$$
$$= +250 \text{ N m}$$

(b) The bending moment due to the part of the beam to the right of the point concerned will be

$$\text{Bending moment} = +250 \times 2.5 - 500 \times 0.5$$
$$= +375 \text{ N m}$$

We could have considered the bending moment due to the part of the beam to the left of the point concerned.

$$\text{Bending moment} = +250 \times 1.5$$
$$= +375 \text{ N m}$$

For this particular beam problem the bending moment obviously depends on which part of the beam we are dealing with.

Example 2 A cantilever of length 3 m carries a load of 2 kN at its free end. How does the bending moment vary along the length of the beam?

$$\text{Bending moment 0.5 m from the free end} = -2 \times 0.5 \text{ kN} \times \text{m}$$
$$= -1 \text{ kN m}$$
$$\text{Bending moment 1.0 m from the free end} = -2 \times 1 \text{ kN} \times \text{m}$$
$$= -2 \text{ kN m}$$
$$\text{Bending moment 1.5 m from the free end} = -2 \times 1.5 \text{ kN} \times \text{m}$$
$$= -3 \text{ kN m}$$

The following table summarises all the results for the bending moments every 0.5 m along the cantilever from the free end.

Distance/m	0	0.5	1.0	1.5	2.0	2.5	3.0
Bending moment/kN m	0	−1	−2	−3	−4	−5	−6

The bending moment increases as the distance from the free end increases, being directly proportional to the distance from the free end. The bending moment is a maximum at the fixed end. All the bending moments are negative, because they cause hogging.

BENDING MOMENT DIAGRAMS FOR BEAMS

Bending moments can be calculated for each section of a bent beam. The variation of the bending moment along the length of a beam depends on the way the beam is loaded and supported. Figure 3.7 shows some examples of such variations. The graphs of bending moment with distance along the beam are called *bending moment diagrams.*

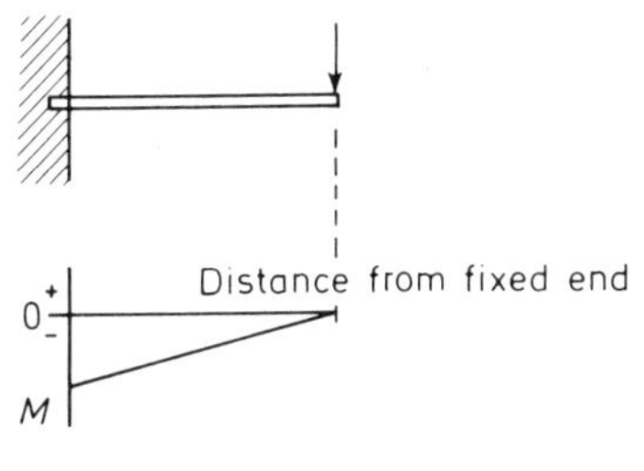

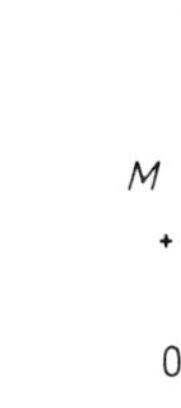

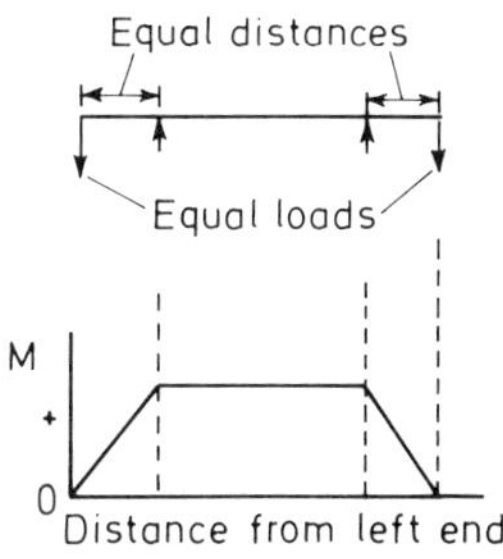

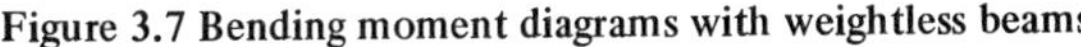

(a) A cantilever

(b) Simply supported beam

(c) The centre part of the beam between the pivots bends into the arc of a circle

Figure 3.7 Bending moment diagrams with weightless beams

For the cantilever the maximum bending moment occurs at the fixed end. For the beam supported at each end and centrally loaded the maximum bending moment occurs at the central loading point.

UNIFORMLY DISTRIBUTED LOADS

The beams so far considered in this chapter have been assumed to have no mass and thus the only loading applied to the beams has been that due to externally applied concentrated loads. The beam's own mass will however contribute a load. The beam might, for instance, sag under its own weight. There could be other loads distributed over the length of the beam.

For a uniformly distributed load, the load can be considered to act at the centroid of the beam. Thus if the gravitational force per unit length of the beam is W then for a length of beam of L the load can be considered to be LW at a distance of $L/2$ from one end. Thus when the bending moment for a particular section is being considered a moment has to be taken into account for the gravitational force WL acting at the centroid of the piece of the beam being considered.

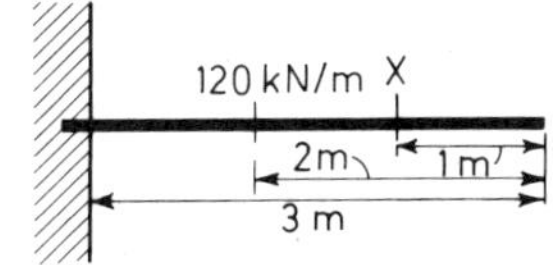

Figure 3.8

Example 3 Figure 3.8 shows a cantilever with a weight per metre length of 120 kN. The cantilever has a length of 3 m and carries no other load. How does the bending moment vary along the length of the beam?

Consider the bending moment for some section X a distance of 1 m from the free end of the cantilever. To the right of the section there is a length of beam of 1 m and this has a weight of 120 kN. This weight can be considered to act at the centroid of that piece and thus be a distance of 0.5 m from X. The bending moment is thus that due to a force of 120 kN a distance of 0.5 m away.

$$\text{Bending moment at X} = -120 \times 0.5 \quad \text{kN} \times \text{m}$$
$$= -60 \text{ kN m}$$

At a distance of 2 m from the free end the weight of the beam to the right will be 240 kN and considered as acting at a distance of 1 m away.

$$\text{Bending moment at Y} = -240 \times 1 \quad \text{kN} \times \text{m}$$
$$= -240 \text{ kN m}$$

The following are the results for other distances from the free end of the beam

Distance/m	0	0.5	1.0	1.5	2.0	2.5	3.0
Bending moment/kN m	0	−15	−60	−135	−240	−375	−540

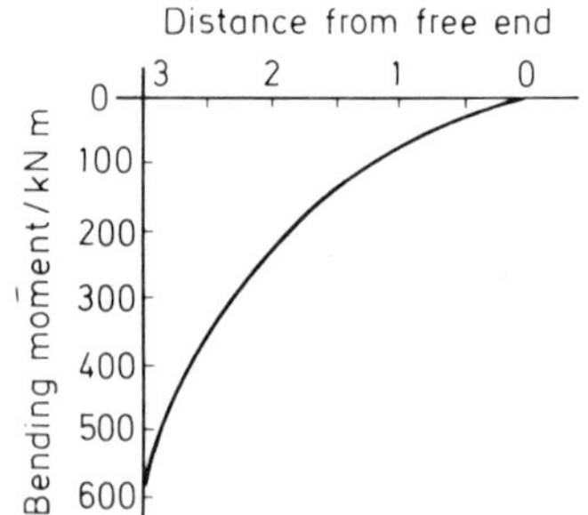

Figure 3.9

The bending moments are all negative because they are tending to make the beam hog.

Figure 3.9 shows how the bending moment for the cantilever varies with distance along the cantilever. The maximum bending moment is at the fixed end.

Example 4 Calculate the maximum bending moment occurring with a 4 m long cantilever of 10 kN/m weight and carrying a load of 8 kN a distance of 3 m from the fixed end.

The distributed load to the right of the fixed end is 4 × 10 kN and can be considered to be located at a distance of 2 m from the fixed end. The maximum bending moment occurs at the fixed end and thus the bending moment due to both the load components is

$$40 \times 2 + 8 \times 3 = 104 \text{ kN m}$$

Because the bending moment is tending to make the beam hog the result is −104 kN m.

BENDING STRESSES

When a beam is bent, as for example in Figure 3.3, stresses are produced. In Figure 3.3 the upper surface of the beam is compressed and the lower surface extended. Stresses produced as a result of bending are called *bending stresses.*

A simple form of bending occurs when a beam is bent to form the arc of a circle. This can occur, for example, when the bending is as in Figure 3.7(c). When the bending moments are applied the beam bends into the arc of a circle with the neutral axis having a radius R (Figure 3.10). If the neutral axis for the part of the beam considered has a length PQ then as

Arc length = radius × angle subtended

$PQ = R\theta$

The neutral axis does not change in length when the beam bends and thus PQ is the length both when bent and before bending. It is thus the length of the layer ST before it bends.

Initial length of $ST = PQ = R\theta$

Length when bent $ST = (R + y)\theta$

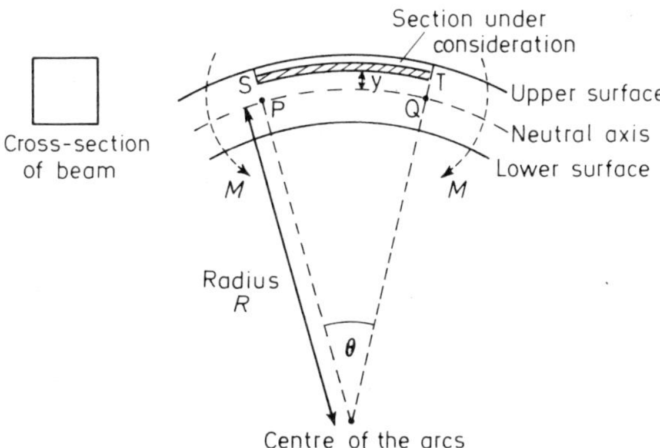

Figure 3.10

The arc ST has a radius of $(R + y)$. The layer ST thus changes in length from $R\theta$ to $(R + y)\theta$ when bending occurs. The change in length is therefore $y\theta$. The strain for the layer ST when bent is

$$\text{Strain} = \frac{\text{change in length}}{\text{original length}} = \frac{y\theta}{R\theta} = \frac{y}{R}$$

If Hooke's law is assumed to be obeyed then

Stress σ = strain × modulus of elasticity E

and therefore the stress acting on the layer ST when bent is

$$\sigma = \frac{yE}{R}$$

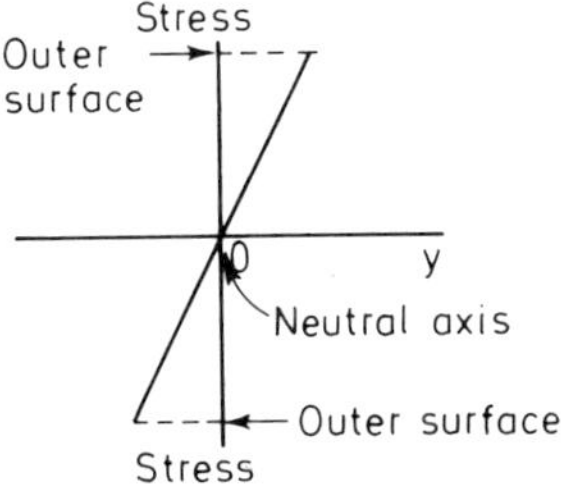

Figure 3.11

The stress on any layer thus depends on its distance from the neutral axis y, the further the layer is from the neutral axis the greater the stress. The stress also depends on the radius to which the beam is bent, the smaller the radius the greater the stress.

The maximum stress for a bent beam occurs at the maximum distance from the neutral axis. Hence the maximum stress will be at the outer surface of the material, furthest from the neutral axis. As the stress is proportional to the distance from the neutral axis then a graph of stress with this distance is a straight line. Figure 3.11 shows such a graph.

POSITION OF THE NEUTRAL AXIS

The stress acting on a layer in a beam is given by

$$\sigma = \frac{yE}{R}$$

where y is the distance of the layer from the neutral axis, R the radius to which the beam is bent and E the modulus of elasticity. However, before the equation can be used for stress calculations we need to know the location of the neutral axis, that axis along which the stress is zero.

Figure 3.12 shows the part of the bent beam considered in arriving at the above equation (Figure 3.10). Suppose the stress acting on the layer ST, a distance y from the neutral axis, is σ. The longitudinal forces acting over the cross-sectional area δA of that layer must be

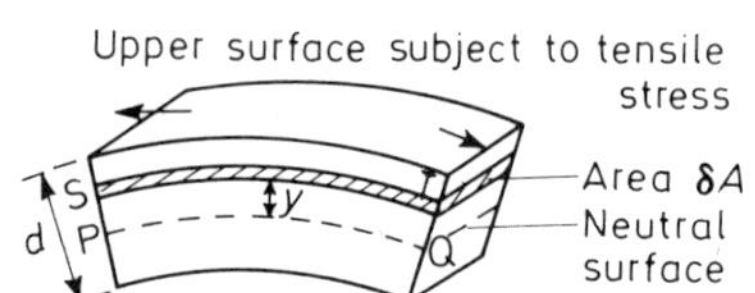

Figure 3.12

$$\text{Force} = \text{stress} \times \text{area}$$
$$= \sigma\delta A$$

Using the relationship for stress given at the beginning of this section gives

$$\text{Force} = \frac{yE}{R}\delta A$$
$$= \frac{E}{R}y\delta A$$

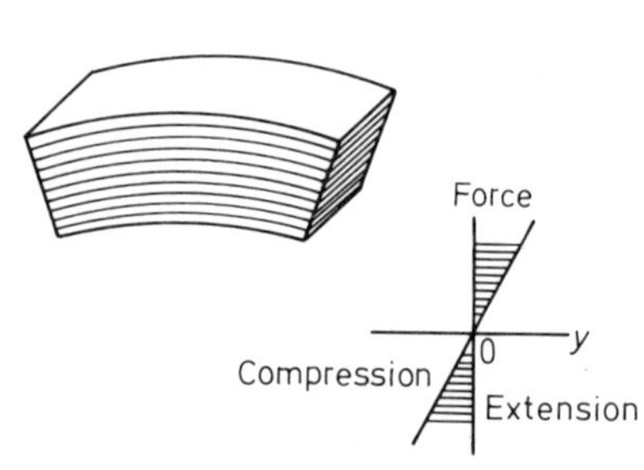

Figure 3.13

This is the force acting on just one layer. The beam is however made up of many layers (Figure 3.13), and each layer has forces acting on it (see Figure 3.11). The total force acting on the beam section must therefore be the sum of all the forces due to each layer.

$$\text{Total force} = \text{sum of each } \frac{E}{R}y\delta A$$

This can be represented mathematically as

$$\text{Total force} = \Sigma \frac{E}{R} y\delta A$$

As E and R are the same for every layer the total force is just a constant E/R multiplied by the sum of all the values of $y\delta A$.

$$\text{Total force} = \frac{E}{R} \Sigma y\delta A$$

If the beam is acted on only by bending moments and has no longitudinal force, i.e. it is only bent, then the total force must be zero. For this to be zero then, as E and R are not zero

$$\Sigma y\delta A = 0$$

But $\Sigma y\delta A$ is the first moment of area (see Chapter 1) about the neutral axis, i.e. the neutral axis from which the distance y is measured. Thus we must have for the pure bending condition that

$$\text{First moment of area} = 0$$

This is only the case if the first moment of area has been considered about the centroid. Thus the *neutral axis must pass through the centroid of the beam section*. This is true regardless of the shape of the beam cross-section.

In the case of a rectangular cross-section beam this means that the neutral axis passes along the centre of the beam. If the beam has a thickness of d then the stress acting on the surface, a distance of $d/2$ from the neutral axis, is

$$\text{Stress on surface} = \frac{E}{R} \times \frac{d}{2}$$

Example 5 Calculate the smallest radius to which a steel strip of thickness 4 mm can be bent if the maximum stress acting on the material is not to exceed 100 N/mm^2. The material has a modulus of elasticity of 210 kN/mm^2.

The maximum stress will occur at the greatest distance from the neutral axis, the surface. Hence

$$\text{Stress } \sigma = \frac{E}{R} \times \frac{d}{2}$$

$$R = \frac{E \times d}{2 \times \sigma}$$

$$= \frac{210 \times 4}{2 \times 100} \qquad \frac{\text{kN/mm}^2 \times \text{mm}}{\text{N/mm}^2}$$

$$= 4.2 \text{ m}$$

Example 6 Calculate the maximum stress that will be produced in a steel beam of rectangular cross-section when bent into the arc of a circle of radius 6 m. The depth of the beam is 6 mm, the modulus of elasticity 210 kN/mm^2.

$$\text{Stress on surface} = \frac{E}{R} \times \frac{d}{2}$$

$$= \frac{210 \times 6}{6000 \times 2} \qquad \frac{\text{kN/mm}^2 \times \text{mm}}{\text{mm}}$$

$$= 105 \text{ kN/mm}^2$$

THE GENERAL BENDING FORMULA

When a beam is bent, there are forces acting on each layer of material within that beam (Figure 3.14). The force acting on a layer a distance y from the neutral axis is given by

$$\text{Force} = \frac{E}{R} y \delta A$$

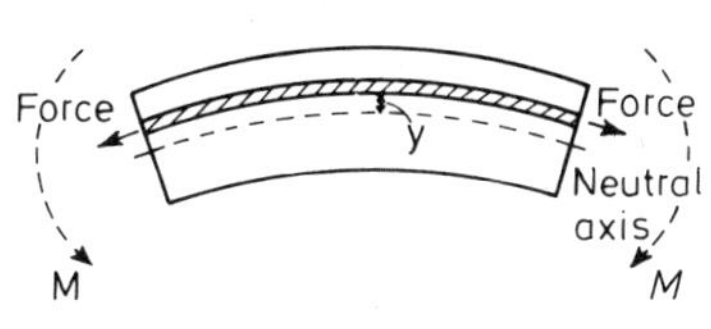

Figure 3.14

This equation was derived in the previous section of this chapter.

The force is acting at a distance y from the neutral axis and so gives a moment. The moment of this force about the neutral axis is

$$\text{Moment of layer} = \text{force} \times y$$

$$= \frac{E}{R} y \delta A \times y$$

$$= \frac{E}{R} y^2 \delta A$$

Each layer (Figure 3.13) will exert a moment about the neutral axis and thus the total moment is the sum of all the moments due to each layer.

$$\text{Total moment} = \text{sum of all } \frac{E}{R} y^2 \delta A$$

This can be expressed as

$$\text{Total moment} = \Sigma \frac{E}{R} y^2 \delta A$$

As E and R are the same for each layer,

$$\text{Total moment} = \frac{E}{R} \Sigma y^2 \delta A$$

The quantity $\Sigma y^2 \delta A$ is called the second moment of area of cross-section and denoted by the symbol I. Hence the equation can be written as

$$\text{Total moment} = \frac{EI}{R}$$

At equilibrium the total moment is equal to the bending moment. Thus

$$M = \frac{EI}{R}$$

or

$$\frac{M}{I} = \frac{E}{R}$$

The stress acting on a layer a distance y from the neutral axis is given by

$$\sigma = \frac{yE}{R}$$

or

$$\frac{\sigma}{y} = \frac{E}{R}$$

This equation and the bending moment equation together give what is called the *general bending formula.*

$$\frac{M}{I} = \frac{\sigma}{y} = \frac{E}{R}$$

The above equation was derived for pure bending where there is only a constant bending moment and no shear or resultant longitudinal forces acting along the beam. A more rigorous analysis is needed to produce an equation for bending where varying bending moments occur and where other forces are involved. The above expression can, however, be used with reasonable accuracy in many cases where these other forces occur. In most cases the maximum bending moment occurs where the shearing force is zero and it is generally a calculation of the maximum stress, which involves the maximum bending moment, that is required.

The value of the second moment of area I about the neutral axis depends on the shape of the beam section concerned. Figure 3.15 shows the values for some common cross-sectional shapes of beams. The units of the second moment of area are $(\text{length})^4$. If the length is in metres then the unit is m^4.

The bending stress is a maximum at the most extreme value of y that is possible. Thus in the case of the rectangular cross-section shown

Neutral axis

$I = \frac{bd^3}{12}$

(a) Rectangular cross-section

Neutral axis

$I = \frac{\pi d^4}{64}$

(b) Circular cross-section

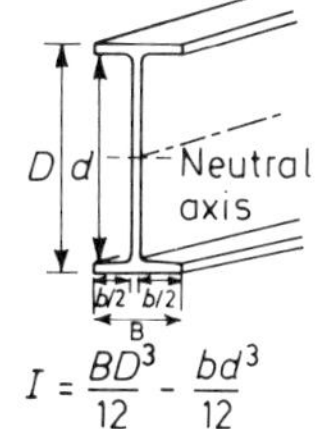

$I = \frac{BD^3}{12} - \frac{bd^3}{12}$

(c) I-cross section

Figure 3.15 Formulae for second moments of area about the neutral axis

in Figure 3.15(a) this is a value for y of $d/2$. The ratio of the bending moment to the maximum stress σ_m is I/y_m where y_m is the maximum possible value of y.

$$\frac{M}{I} = \frac{\sigma}{y}$$

Hence

$$\frac{M}{\sigma_m} = \frac{I}{y_m}$$

The quantity I/y_m is a purely geometrical function, depending only on the shape of the cross-section concerned. In the case of the rectangular cross-section it is given by

$$I = \frac{bd^3}{12}$$

and

$$y_m = \frac{d}{2}$$

thus

$$\frac{I}{y_m} = \frac{bd^3/12}{d/2}$$

$$\frac{I}{y_m} = \frac{bd^2}{6}$$

This quantity is called the *section modulus Z* (or elastic modulus Z) for the rectangular cross-section.

$$Z = \frac{I}{y_m}$$

and thus

$$\frac{M}{\sigma_m} = Z$$

The section modulus is thus the factor that relates the maximum bending stress to the bending moment. The units of the section modulus are $(\text{length})^3$. If the lengths are in metres then the unit will be m^3.

For the beam sections shown in Figure 3.15 the section modulus are:

Rectangular cross-section $Z = \frac{bd^2}{6}$

Circular cross-section $Z = \frac{\pi d^3}{32}$

$$I \text{ cross-section } Z = \frac{BD^3}{12} - \frac{bd^3}{12} \div \frac{D}{2}$$

Standard section handbooks give values of section modulus for different size beams. Such data enables the appropriate beam to be selected for a

Outside diameter /mm	*Thickness* /mm	*Mass* kg/m	*Sectional area* /cm²	*Elastic modulus Z* /cm³
21.3	3.2	1.43	1.82	0.72
26.9	3.2	1.87	2.38	1.27
33.7	2.6	1.99	2.54	1.84
	3.2	2.41	3.07	2.14
	4.0	2.93	3.73	2.49

particular situation. Thus if a tube was required to withstand a particular bending stress when subject to a bending moment the section modulus can be computed and used to determine which size tube should be used. The following apply to universal beams, i.e. I cross-sections. X–X is the horizontal axis (Figure 3.15(c)) and Y–Y is the vertical axis.

Depth of section D	*Width of section B*	*Thickness Webb*	*Thickness Flange*	*Root radius*	*Depth between fillets*	*Area of section*	*Elastic modulus About X–X*	*Elastic modulus About Y–Y*
/mm	/mm	/mm	/mm	/mm	/mm	/cm²	/cm³	/cm³
617.0	230.1	13.1	22.1	12.7	543.1	178.2	3 620	369.6
611.9	229.0	11.9	19.6	12.7	543.1	159.4	3 217	321.1
607.3	228.2	11.2	17.3	12.7	543.1	144.3	2 874	279.1
602.2	227.6	10.6	14.8	12.7	543.1	129.0	2 509	233.6

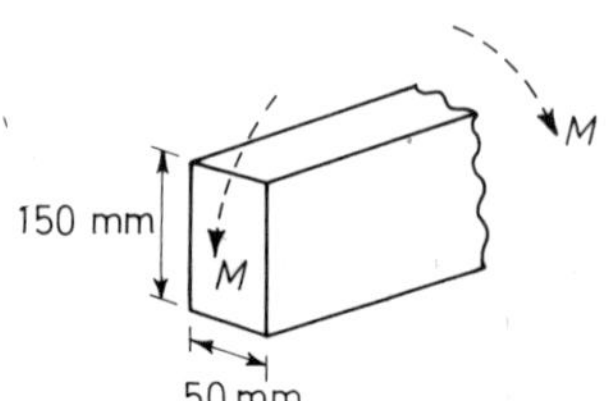

Figure 3.16

Example 7 Determine the maximum stress set up in a rectangular cross-section timber beam, 150 mm by 50 mm, when a bending moment of 500 N m is applied. Figure 3.16 shows the way in which the beam is bent.

$$\sigma_m = \frac{M}{Z} = \frac{500}{\dfrac{0.05 \times 0.15^2}{6}} \qquad \frac{\text{N m}}{\text{m}^3}$$

$$= 2.7 \times 10^6 \text{ N/m}^2$$

$$= 2.7 \text{ N/mm}^2$$

Example 8 An I-section girder has a section modulus of 25×10^{-5} m³. What will be the maximum stress produced when such a beam is subject to a bending moment of 30 kN m?

$$\sigma_m = \frac{M}{Z} = \frac{30 \times 10^3}{25 \times 10^{-5}} \qquad \frac{\text{N m}}{\text{m}^3}$$

$$= 120 \times 10^6 \text{ N/m}^2$$

$$= 120 \text{ N/mm}^2$$

Example 9 Determine the maximum bending moment that can be applied to a bar with a circular cross-section of diameter 50 mm if the maximum bending stress is not to exceed 120 N/mm².

$$Z = \frac{\pi d^3}{32} = \frac{\pi \times 0.05^3}{32} \qquad \text{m}^3$$

$$M = Z\sigma_\text{m} = \frac{\pi \times 0.05^3}{32} \times 120 \times 10^6 \qquad \text{m}^3 \times \text{N/m}^2$$

$$= 1.5 \times 10^3 \text{ N m}$$

Example 10 A uniform I-section girder 6 m long is supported at its ends. Loads of 40 kN are carried at 1.5 m from each end. Calculate the maximum bending stress. The section modulus of the beam is 4.0×10^{-3} m³

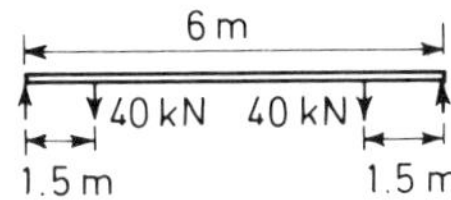

Figure 3.17

Figure 3.17 shows the general arrangement of the beam. The reactions at each end of the beam will be the same, the beam being symmetrically loaded. Hence, as the upwards directed forces equal the downwards directed forces, each reactive force must be 40 kN. If you are not sure of this, take moments about a point and solve the simultaneous equations given by the moment equation and the equilibrium equation produced for the forces in the vertical direction.

The bending moment for any position between the two loading positions is a constant. Try taking the moment about any point in that region. For the moment to the left of the central point of the beam:

$$40 \times 3.0 - 40 \times 1.5 = 6.0 \text{ kN m}$$

$$\sigma_\text{m} = \frac{M}{Z} = \frac{6.0 \times 10^3}{4.0 \times 10^{-3}} \qquad \frac{\text{N m}}{\text{m}^3}$$

$$= 1.5 \times 10^6 \text{ N/m}^2$$

$$= 1.5 \text{ N/mm}^2$$

Example 11 Calculate the maximum bending stress for the I-section girder in the previous question if the girder has a weight of 30 kN/m.

The total weight of the girder is $30 \times 6 = 180$ kN. Thus the total downwards forces are $180 + 40 + 40 = 260$ kN. The reactive forces at the supports will therefore be 130 N.

For the moment to the left of the central point of the beam:

$$130 \times 3.0 - 40 \times 1.5 - 130 \times 1.5 = 135 \text{ kN m}$$

The last term above is that due to the weight of that part of the beam being considered to act at the centroid of that part.

$$\sigma_\text{m} = \frac{M}{Z} = \frac{135 \times 10^3}{4.0 \times 10^{-3}} \qquad \frac{\text{N m}}{\text{m}^3}$$

$$= 34 \times 10^6 \text{ N/m}^2 \text{ to two significant figures.}$$

$$= 34 \text{ N/mm}^2$$

PROBLEMS

1. Explain the terms neutral surface, neutral axis and bending moment.

2. A beam of length 3 m rests on supports at each end. A load of 3 kN is applied to the beam centre. What will be the bending moments a distance of (a) 0.5 m, (b) 1.0 m, (c) 1.5 m from one end? Where will the bending moment be a maximum?

3. A cantilever of length 2.0 m carries a load of 12 kN at its free end. What, and where, will be the maximum bending moment?

4. Draw bending moment diagrams showing how the bending moment varies along the length of the beams shown in Figure 3.18.

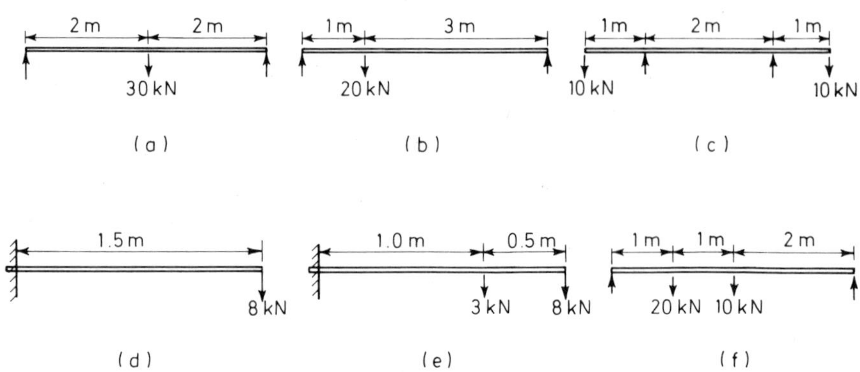

Figure 3.18

5. Draw bending moment diagrams for each of the beams shown in Figure 3.18 if each beam has a weight of 10 kN/m.

6. Calculate the maximum bending moment for a cantilever having a weight of 20 kN/m and a length of 3 m.

7. Steel strip of thickness 5 mm is coiled on a drum of 1.5 m diameter. Calculate the maximum stress produced by the coiling. The modulus of elasticity is 210 kN/mm^2.

8. Calculate the minimum diameter a drum may have for aluminium strip to be wound on if the maximum stress due to the coiling must not exceed 100 N/mm^2. The strip has a thickness of 3 mm. The modulus of elasticity is 70 kN/mm^2.

9. A strip of steel 2 mm thick passes over a pulley. Determine the diameter of the pulley needed if the stress in the steel strip must not exceed 100 N/mm^2 as a result of it bending to follow the circumference of the pulley wheel. The steel has a modulus of elasticity of 210 kN/mm^2.

10. Calculate the maximum stresses that will be set up in a rectangular cross-section timber beam, 200 mm by 50 mm, when a bending moment of 500 N m is applied to it when the beam bends with (a) its smaller face, (b) its larger face parallel to the neutral surface.

11. An I-section girder has a section modulus of 100×10^{-5} m^3. What will be the maximum stresses produced when such a beam is subject to a bending moment of 40 kN m?

12. A steel strip 3 mm thick and 50 mm wide is to bend round a drum of diameter 4.0 m. What bending moment is required to bend the strip round the drum and what will be the maximum stresses produced? The modulus of elasticity of the steel is 210 kN/mm^2.

13. A wooden joist of rectangular cross-section spans a gap of 5 m. The joist has a depth of 200 mm. Calculate the minimum width of joist required if the maximum tensile stress that the beam can be allowed to withstand is 7 N/mm^2 and the load to be carried is a centrally placed load of 20 kN.

14. An axle of diameter 120 mm and length 3.0 m is supported symmetrically and horizontally between two bearings 1.5 m apart. What is the maximum stress produced in the bearings if vertical loads of 20 kN are applied at each end of the axle?

15. A timber beam of thickness 100 mm and depth 300 mm protrudes 2 m from a wall in which it is fixed. What is the maximum load that can be applied to the free end of the beam if the maximum allowable stress is 7 N/mm^2?

16. What would be the maximum load that the timber beam in Problem 15 could carry if the load was not located at the free end but uniformly distributed over the entire length of the protruding beam?

17. Calculate the maximum uniformly distributed load that a simply supported steel I-section can carry over a span of 6 m if the maximum permissible stress in the beam is 50 N/mm^2 and the I-section has a section modulus of 4×10^{-3} m^3.

18. An I-section girder is 140 mm deep with flanges 100 mm wide by 12 mm thick and a web thickness of 8 mm. (a) What is the section modulus? (b) A 6 m long piece of the above girder is simply supported at its ends and loads of 50 kN are applied at distances of 1.5 m from each end. What is the maximum bending moment? (c) What is the maximum stress? (d) What is the radius of curvature of the central part of the beam? The modulus of elasticity is 210 kN/mm^2.

19. A metal rod has a rectangular section 120 mm deep and 40 mm wide. One end of the metal rod is rigidly built in. Calculate the maximum load that can be applied to the other end of the rod if the stress is not to exceed 15 N/mm^2 and the protruding length of the rod is 200 mm.

20. A simple wooden bridge consists of four parallel timber beams, each 250 mm deep and 100 mm wide, spanning a gap of 3 m. Planks rest on the beams to provide the walk-way. If the maximum allowable stress for a beam is 7 N/mm^2 what is the maximum uniformly distributed load that the bridge can bear.

21. For this problem, use the data given for circular hollow sections in the chapter. (a) What will be the maximum bending stress produced for a circular hollow steel section of outside diameter 21.3 mm and wall thickness 3.2 mm when subject to a bending moment of 300 N m? (b) What is the smallest mass per metre steel circular hollow section that can be used if the maximum bending stress is not to exceed 100 N/mm^2 when the bending moment is 200 N m?

22. Select from the data given in this chapter for I-sections a beam for which the maximum bending stress will be 75 N/mm^2 when the bending moment is 200 kN m. The bending is about the X–X axis.

23. For the circular hollow sections specified in this chapter the steel for a particular grade has a tensile strength of 430/540 N/mm^2 and a minimum yield stress of 255 N/mm^2. Propose a design for a child's climbing frame based on the use of one of these circular sections.

4 Torsion

TWISTING SHAFTS

A motor is supplied with power and causes a shaft to rotate. At the end of the shaft is a pulley wheel (Figure 4.1). The rotating shaft causes the pulley to rotate and lift a load. The load gains energy. The power input to the motor has thus been transmitted by the shaft to the load. But how is the power transmitted along the shaft?

If a strip of rubber is held with one end in each hand, then one end twisted, a twisting action can be seen, and felt, being transmitted along the rubber. The twisting action communicated to one end of the rubber by one hand is transmitted along the rubber and results in a twisting action being communicated to the other hand. The power is transmitted along the rubber by the rubber twisting.

The shaft of the motor behaves in the same way. The motor exerts a twisting action on one end of the shaft. This is then transmitted along the shaft and results in the twisting action being communicated to the pulley wheel. This twisting action is called *torsion.*

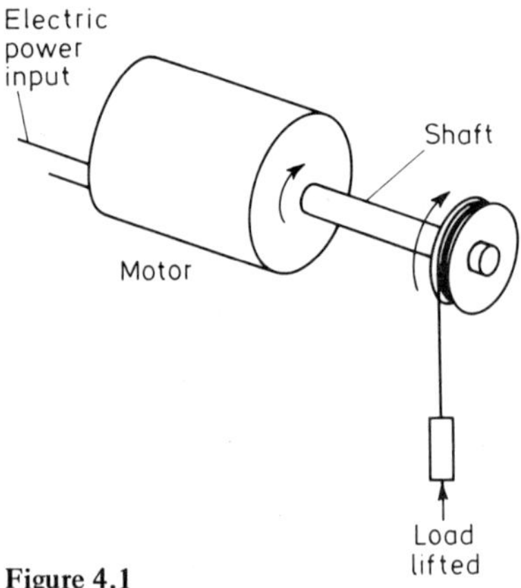

Figure 4.1

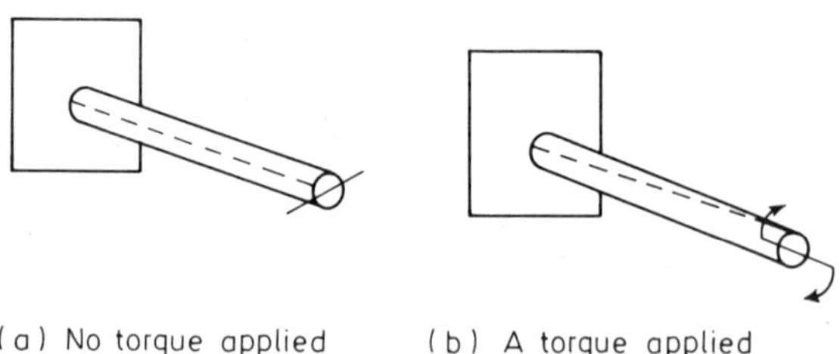

Figure 4.2 Twisting one end of a shaft

Figure 4.2(a) shows a shaft held rigidly at one end. No torque is applied to the shaft and thus a mark longitudinally along the surface of the bar is a straight line parallel to the axis of the bar. When, however, a torque is applied, Figure 4.2(b), the bar twists and the mark becomes distorted. The further down the bar, from the end at which the torque was applied, the less the movement of the mark from its initial position. The twisting action has however been transmitted down the bar. Torque has been experienced by the different sections of the bar.

TORSIONAL STRESSES

When a circular shaft is subject only to a torque centred on its longitudinal axis then lines which are radial before twisting remain as radial after twisting (Figure 4.3). This means that all circular sections of the shaft remain circular during the twisting with diameters unchanged. When the above occurs the twisting is said to be *pure torsion*. This chapter is only concerned with pure torsion.

When the shaft is twisted each small area of the surface of the shaft becomes distorted. Figure 4.4 shows what happens. A square shape on the surface becomes sheared (see Chapter 2). The shear strain is ϕ. If the shear stress is τ then, to the limit of proportionality,

Figure 4.3 The twisting of each section of a shaft when subject to torque. Note that each section remains circular

$$\text{modulus of rigidity } G = \frac{\text{shear stress}}{\text{shear strain}}$$

$$G = \frac{\tau}{\phi} \text{ and so } \phi = \frac{\tau}{G}$$

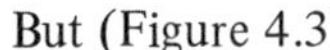

But (Figure 4.3)

$$\text{arc } AB = L\phi \qquad \text{if } \phi \text{ is small.}$$

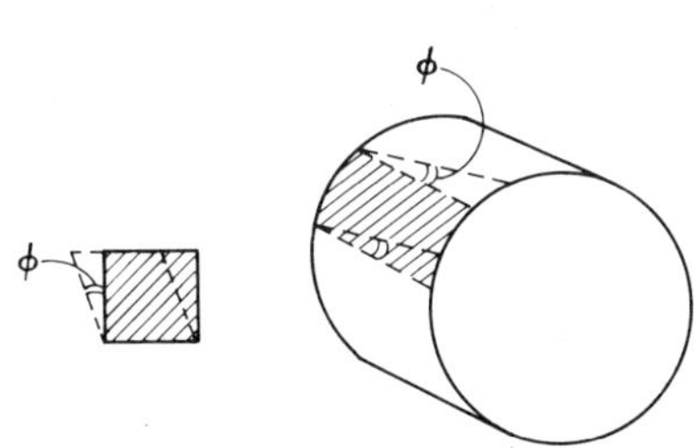

Figure 4.4 The change in the shape of a surface element when a shaft is subject to torque

(The arc of a circle is the radius multiplied by the angle subtended.) Also (Figure 4.3)

$$\text{arc } AB = r\theta$$

where θ is the angle by which a length of shaft L is twisted. By combining the two equations we have

$$L\phi = r\theta$$

$$\phi = \frac{r\theta}{L}$$

Thus

$$\text{shear strain} = \frac{r\theta}{L} = \frac{\tau}{G} \qquad (4.1)$$

This is the shear strain at the surface of the shaft of radius r. The equation is also valid for the shear strain, or shear stress, at a ring of radius r within a shaft of greater radius than r. The expression indicates how the shear strain, and shear stress, vary with radial distance from the centre of the shaft. The maximum value is obviously at the maximum value of r, i.e. the surface, the shear strain and shear stress being proportional to the radial distance (Figure 4.5).

The shaft can be considered to be made up of a number of thin concentric tubes. As each such tube has a different radius there will be different shear strain and shear stress for each. Figure 4.6(a) shows the concept of the concentric tubes and Figure 4.6(b) shows the cross-section of just one of the tubes. The tube has a radius of r_1 and a thickness of δr_1. The cross-sectional area of the tube is thus

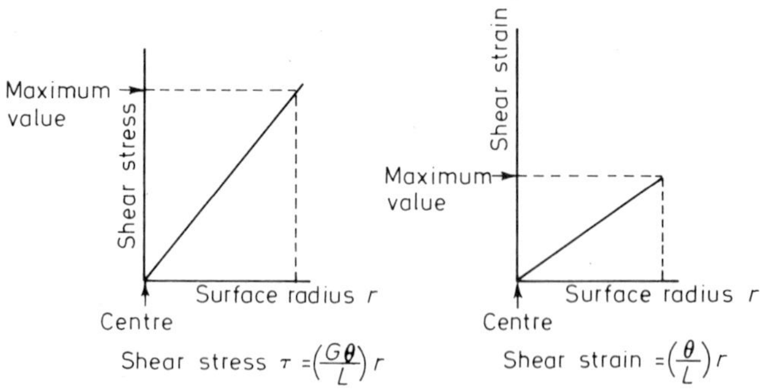

Figure 4.5

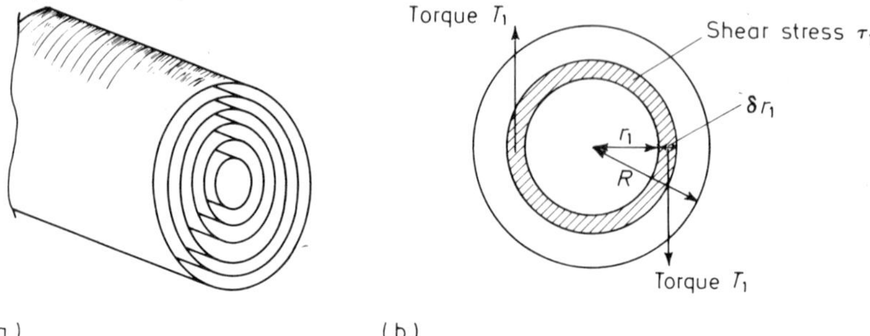

Figure 4.6 (a) The shaft considered as a series of concentric tubes. (b) The torque acting on one of the concentric tubes

$$\text{Cross-sectional area} = 2\pi r_1\,\delta r_1$$

(The area is the circumference multiplied by the thickness of the tube.) If at that radius the shear stress is τ_1 then

$$\text{Shear stress} = \frac{\text{shearing force}}{\text{area}}$$

and so

$$\text{Shearing force} = \tau_1 \times 2\pi r_1\,\delta r_1$$

The moment of this force about the shaft axis is

$$\text{Moment} = \tau_1 \times 2\pi r_1\,\delta r_1 \times r_1$$

The equation developed earlier in this section, Equation (4.1), gives a relationship between shear stress and radius. Hence the shear stress τ_1 at a radius r_1 is given by

$$\frac{r_1\theta}{L} = \frac{\tau_1}{G}$$

Thus the moment equation can be written as

$$\text{Moment} = \frac{r_1\theta G}{L} \times 2\pi r_1\,\delta r_1 \times r_1$$

$$= \frac{G\theta}{L} \times 2\pi r_1^3\,\delta r_1$$

This is the moment of the shearing force for just one of the tubes which make up the shaft. Each tube will have a different moment because it has a different radius and force. The total moment, i.e. torque, on the shaft will be given by the sum of the moments of the shearing forces on all the tubes that make up the shaft.

$$\text{Total torque, } T = \Sigma \frac{G\theta}{L} \times 2\pi r^3 \, \delta r$$

$$T = \frac{G\theta}{L} \times 2\pi \Sigma r^3 \, \delta r$$

Only r is varying for the different tubes. The quantity $2\pi\Sigma r^3 \, \delta r$ is called the *polar moment of area* J of the cross-section about the shaft axis. Thus

$$T = \frac{G\theta J}{L}$$

The above equation can be combined with Equation (4.1) to give a *general equation for the torsion* of circular cross-section shafts.

$$\frac{T}{J} = \frac{G\theta}{L} = \frac{\tau}{r}$$

The polar moment of area J is given by

$$J = 2\pi \Sigma r^3 \, \delta r$$

If very small values of δr are considered the summation can be written as an integral.

$$J = 2\pi \int_0^R r^3 \, \mathrm{d}r$$

$$J = \frac{\pi R^4}{2}$$

This is the value of the polar moment for a solid shaft. If the shaft was hollow, having an internal radius of r and an external radius R then

$$J = 2\pi \int_r^R r^3 \, \mathrm{d}r$$

$$J = \frac{\pi}{2}(R^4 - r^4)$$

If diameters are used instead of radii then as, for the solid shaft, $R = D/2$ with D being the diameter

$$\text{Solid shaft } J = \frac{\pi D^4}{32}$$

and for the hollow shaft $R = D/2$ and $r = d/2$ with D being the external diameter and d the internal diameter,

$$\text{Hollow shaft } J = \frac{\pi}{32}(D^4 - d^4)$$

The units of J are m^4 if the radii or diameters are in m.

For a solid shaft of radius R the general equation for torsion gives

$$\frac{T}{J} = \frac{\tau}{R}$$

or

$$T = \frac{J}{R} \times \tau$$

Both J and R are purely geometrical functions, only involving the shaft radius. The quantity J/R can be replaced by a quantity Z_p called the *polar modulus of section.*

$$Z_p = \frac{J}{R}$$

For a solid shaft Z_p will have the value $\pi R^3/2$. Thus

$$T = Z_p \tau$$

The units of Z_p are m^3 if the radii or diameters are in m.

Example 1 A torque of 30 kN m is to be transmitted by a solid shaft. Calculate the diameter of the shaft if the shear stress is not to exceed 80 MN/m² (i.e. 80 N/mm²).

$$T = \frac{J}{R} \times \tau$$

For a solid shaft

$$\frac{J}{R} = \frac{\pi R^3}{2}$$

Hence

$$T = \frac{\pi R^3}{2} \times \tau$$

Hence

$$R^3 = \frac{2T}{\pi\tau}$$

$$= \frac{2 \times 30 \times 10^3}{\pi \times 80 \times 10^6} \qquad \frac{\text{N m}}{\text{N/m}^2} = \text{m}^3$$

$$R = 62 \times 10^{-3} \text{ m to two significant figures}$$

Hence the diameter is 124×10^{-3} m or 124 mm.

Example 2 Calculate the external diameter of a hollow shaft needed to transmit a torque of 30 kN if the shear stress is not to exceed 80 MN/m² and the shaft has an external diameter twice the internal diameter.

$$T = \frac{J}{R} \times \tau$$

For a hollow shaft

$$\frac{J}{R} = \frac{\pi}{2R}(R^4 - r^4)$$

Hence as $R = 2r$

$$\frac{J}{R} = \frac{\pi}{2R}\left\{R^4 - \left(\frac{R}{2}\right)^4\right\}$$

$$\frac{J}{R} = \frac{\pi R^3}{2} \times \frac{7}{8}$$

Thus

$$T = \frac{7\pi R^3}{16} \times \tau$$

and so

$$R^3 = \frac{16T}{7\pi\tau}$$

$$= \frac{16 \times 30 \times 10^3}{7 \times \pi \times 80 \times 10^6} \qquad \frac{\text{N m}}{\text{N/m}^2} = \text{m}^3$$

$$R = 108 \times 10^{-3} \text{ m to three significant figures}$$

Hence the diameter is 216×10^{-3} m or 216 mm.

Example 3 Calculate the maximum torque that can be transmitted by a solid shaft of diameter 120 mm if the shearing stress is not to exceed 80 MN/m^2 (i.e. 80 N/mm^2).

$$T = \frac{J}{R} \times \tau$$

For a solid shaft

$$\frac{J}{R} = \frac{\pi R^3}{2}$$

Hence

$$T = \frac{\pi R^3}{2} \times \tau$$

$$= \frac{\pi \times (60 \times 10^{-3})^3 \times 80 \times 10^6}{2} \qquad \text{m}^3 \times \frac{\text{N}}{\text{m}^2} = \text{Nm}$$

$$= 27.1 \times 10^3 \text{ N m to three significant figures}$$

$$= 27.1 \text{ kN m}$$

Example 4 Calculate the maximum shear stress produced for a 6 mm diameter bolt when it is tightened by a spanner which applies a torque of 7 N m.

$$T = \frac{J}{R} \times \tau$$

For a solid shaft

$$\frac{J}{R} = \frac{\pi D^3}{16}$$

Hence

$$T = \frac{\pi D^3 \times \tau}{16}$$

and so

$$\tau = \frac{16T}{\pi D^3}$$

$$= \frac{16 \times 7}{\pi \times (6 \times 10^{-3})^3} \qquad \frac{\text{N m}}{\text{m}^3}$$

$$= 170 \times 10^6 \text{ N/m}^2 \text{ to two significant figures}$$

$$= 170 \text{ N/mm}^2$$

TRANSMISSION OF POWER

Power is the rate of transfer of energy. With the transfer occurring by means of work then the power becomes the rate at which work is done. The *work* done by the point of application of a force F moving through a distance s in the line of action of the force is given by

$$\text{Work} = F \times s$$

The unit of work is the watt (W).

For a rotating body with a torque T applied at a distance r from the pivot point then the distance travelled by the point of application of the force is given by

$$\text{Distance travelled for one revolution} = 2\pi r$$

and if n revolutions are made per second then the distance travelled per second is

$$\text{Distance travelled per second} = 2\pi rn$$

The work done per second will be

$$\text{Work done per second} = 2\pi rnF$$

$$= 2\pi n \times Fr$$

But Fr is the torque T, thus

$$\text{Work done per second} = 2\pi nT$$

But, the work done per second is the power. Thus

$$\text{Power} = 2\pi nT$$

This is the power transmitted.

The *angular velocity* ω is the rate at which a point on the object sweeps out the angle at the centre per second. There are 2π radians in one complete revolution, thus if n revolutions are made per second the angle swept out per second is $2\pi n$. Thus

$$\omega = 2\pi n$$

Hence the power transmitted can be written as

$$\text{Power} = \omega T$$

If the units of ω are radian/s (or just s^{-1}), and the torque T has units of newton metre, the power is given in watts.

Example 5 Calculate the power that can be transmitted by a solid steel shaft of diameter 100 mm rotating at 5 revolutions per second if the shear stress must not exceed 70 MN/m^2 (i.e. 70 N/mm^2).

$$\text{Power} = 2\pi nT$$

$$T = \frac{J\tau}{R}$$

Hence

$$\text{Power} = 2\pi n \frac{J}{R} \tau$$

For a solid shaft

$$\frac{J}{R} = \frac{\pi R^3}{2}$$

Hence

$$\text{Power} = 2\pi n \frac{\pi R^3}{2} \tau$$

$$= \pi^2 \times 5 \times (50 \times 10^{-3})^3 \times 70 \times 10^6 \text{ s}^{-1} \times \text{m}^3 \times \text{N/m}^2$$

$$= 432 \times 10^3 \text{ N m s}^{-1} \text{ to three significant figures}$$

$$= 432 \times 10^3 \text{ W}$$

$$= 432 \text{ kW}$$

Example 6 The drive shaft of a car is a hollow tube with an external diameter of 50 mm and an internal diameter of 47 mm. What is the shear stress when the shaft is transmitting a power of 70 kW and rotating at 80 revolutions per second?

$$\text{Power} = 2\pi nT$$

$$T = \frac{J}{R}\tau$$

Hence

$$\text{Power} = 2\pi n \frac{J}{R} \tau$$

For a hollow shaft

$$\frac{J}{R} = \frac{\pi}{32R}(D^4 - d^4)$$

Thus

$$\text{Power } P = 2\pi n \frac{\pi}{32R}(D^4 - d^4)\tau$$

and so

$$\tau = \frac{16RP}{\pi^2 n(D^4 - d^4)}$$

$$= \frac{16 \times 50 \times 10^{-3} \times 70 \times 10^3}{\pi^2 \times 80\{(50 \times 10^{-3})^4 - (47 \times 10^{-3})^4\}}$$

$$\frac{\text{m} \times \text{W}}{\text{s}^{-1} \times \text{m}^4} = \frac{\text{m} \times \text{N} \times \text{m} \times \text{s}^{-1}}{\text{s}^{-1} \times \text{m}^4}$$

$$= 51 \times 10^6 \text{ N/m}^2 \text{ to 2 significant figures}$$

$$= 51 \text{ N/mm}^2$$

PROBLEMS

1. A solid circular steel shaft is subject to a torque of 1 kN m. What will be the diameter of the shaft if the shear stress can be allowed as high as 70 MN/m^2 (i.e. 70 N/mm^2)?

2. Calculate the torque which can be transmitted by a solid circular shaft of diameter 130 mm at a maximum shearing stress of 80 MN/m^2 (i.e. 80 N/mm^2).

3. Calculate the external diameter of a hollow shaft which is needed to transmit a torque of 20 kN m and not have shear stresses exceeding 80 MN/m^2 (i.e. 80 N/mm^2). The ratio of the external and internal diameters is 1.4.

4. Calculate the angle of twist produced in a solid circular shaft of diameter 40 mm and length 1 m when the shear stress is 40 MN/m^2 (i.e. 40 N/mm^2) and the shear modulus 80 GN/m^2 (i.e. 80 kN/mm^2).

5. Calculate the maximum shear stress produced in a bolt of diameter 20 mm when it is tightened by a spanner which exerts a force of 60 N with a radius of action of 150 mm.

6. Two sections of a shaft are connected together, on the same longitudinal axis, by a flange coupling which has 6 bolts on a 200 mm pitch circle. If the shaft transmits a torque of 30 kN m, and the allowable shear stress for the bolts is 80 MN/m^2 (i.e. 80 N/mm^2), what will be the required diameter for the bolts?

7. A hollow circular shaft has an internal diameter of 120 mm and an external diameter of 160 mm. Calculate the shear stresses produced at the outer and inner surfaces of the shaft when the applied torque is 40 kN m.

8. A hollow shaft has to carry a torque of 10 kN m and to have an external diameter of 150 mm. Calculate the internal diameter that will be necessary if the maximum shear stress is not to exceed 40 MN/m^2 (i.e. 40 N/mm^2).

9. Calculate the angle of twist produced on a 6 m length of the shaft in Problem 8.

10. Calculate the maximum shear stress in a solid circular shaft of diameter 120 mm when it is transmitting a power of 10 kW at 60 revolutions per minute.

11. A turbine shaft transmits 60 kW at 4 revolutions per second. The shaft has an external diameter of 1 m and a thickness of 25 mm. What is the maximum shear stress produced?

12. Design a solid circular shaft to transmit 100 kW at 180 revolutions per minute with material for which the maximum possible shear stress is 50 MN/m^2 (i.e. 50 N/mm^2).

13. Design a hollow steel shaft to transmit 1.5 MW at 4 revolutions per second. The internal diameter of the shaft is to be 0.7 times the external diameter and the maximum shear stress must not exceed 60 MN/m^2 (i.e. 60 N/mm^2). Calculate the internal and external diameters of the shaft and the angle of twist over a length of 3 m when the above power is being transmitted. The modulus of rigidity of the material can be taken as 80 GN/m^2 (i.e. 80 kN/mm^2).

14. The specification for a shaft states that it must be a solid circular steel shaft capable of transmitting 500 kW when rotating at 240 revolutions per minute. The shear stress must not exceed 50 MN/m^2 (i.e. 50 N/mm^2) and the angle of twist must not exceed one degree in a metre length of shaft. Calculate the diameter of the shaft that will meet the specification. The steel has a modulus of rigidity of 80 GN/m^2 (i.e. 80 kN/mm^2).

15. A coupling between two shafts has to transmit 300 kW at 120 revolutions per minute. Calculate the diameter of the bolts required if six bolts are used at a pitch circle of 300 mm diameter and the shear stress in the bolts must not exceed 70 MN/m^2 (i.e. 70 N/mm^2).

16. The two parts of a solid steel circular shaft each have a diameter of 100 mm. The two parts are coupled together by 8 bolts equally placed on a pitch circle of 200 mm diameter. What should be the diameter of the bolts used if the maximum permissible shear stress for the shaft is 60 MN/m^2 (i.e. 60 N/mm^2) and that for the bolts 80 MN/m^2 (i.e. 80 N/mm^2).

5 Force and motion

NEWTON'S LAWS

When there is no resultant force acting on an object, the object either remains at rest or moves with a constant velocity. This is a statement of *Newton's first law of motion*. The law is a statement of what happens when no resultant force is acting and by implication is stating that if the above is not the case then there must be a resultant force acting.

A large packing case may remain at rest on the factory floor even though you are pushing and convinced that you are exerting a force on it. But, because it is at rest there can be no resultant force acting on the object. Your force must be exactly opposed by an equal force so that there is zero resultant. For the packing case, this opposing force is probably a frictional force, although there could be somebody round the other side of the packing case pushing in the opposite direction to you.

When an object accelerates we say that a resultant force is acting and that it is acting in the direction of the acceleration. We also state that the force is directly proportional to the acceleration. Thus if the acceleration of an object is doubled then the applied resultant force must have doubled. Thus, if the object being accelerated was, say, a standard size building brick then we might have an acceleration of 1 m/s^2 produced by a force F. To obtain an acceleration of 2 m/s^2 would require twice the force, i.e. $2F$. Suppose however we had changed the object to two standard size bricks. The $2F$ force gives an acceleration of just 1 m/s^2. The acceleration that is produced when a force acts on an object depends not only on the size of the force but also a quantity associated with the object itself. This quantity which represents the difficulty with which the object will accelerate, its *inertia*, is called its *mass*. An object which requires twice the force to produce the same acceleration has twice the mass.

The above arguments can be combined to give a version of *Newton's second law of motion*: The product of the mass and acceleration of a body is proportional to the force causing the acceleration, the direction of the acceleration being in the same direction as that of the force. This can be represented by an equation

$$F = ma$$

where F is the resultant force, m the mass and a the acceleration. With units of acceleration as m/s^2 and the mass in kg, defined in terms of a standard block of matter, then the unit of force is defined as being in newtons (N).

Any force is just one aspect of an interaction between two bodies. *Newton's third law of motion* concerns this interaction and can be stated as: if one body exerts a force on a second body then the second exerts an equal and opposite force on the first. This is often written as: to every action there is always opposed an equal reaction. Thus, if you pull on one end of a rope then the rope pulls on you. If you consider the rope to be like a strip of rubber then, when you pull on it, stretching occurs. The rope in trying to contract back to its original length pulls on you, a force in the opposite direction to the force you are applying to cause the stretching.

Example 1 Calculate the resultant force needed to cause a mass of 3 kg to accelerate at 4 m/s^2.

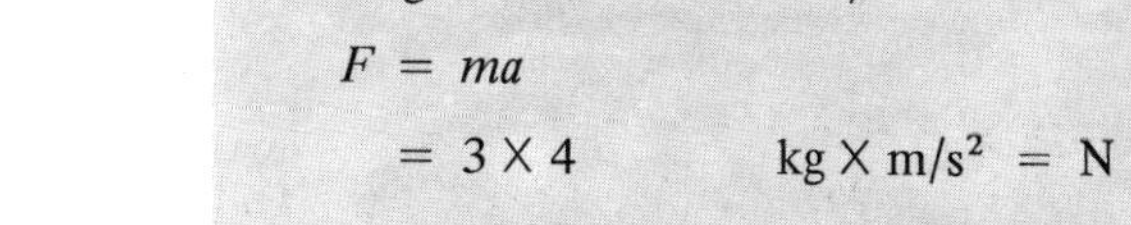

$$F = ma$$

$$= 3 \times 4 \qquad \text{kg} \times \text{m/s}^2 = \text{N}$$

$$= 12\ \text{N}$$

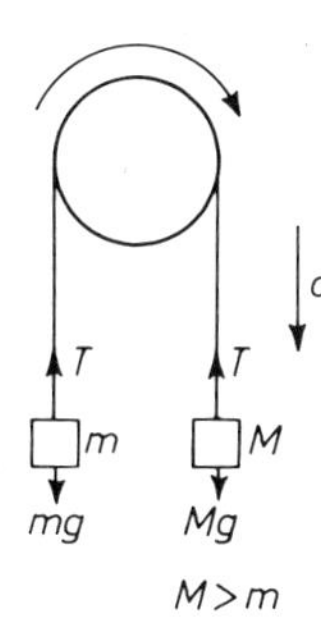

Figure 5.1

Example 2 Figure 5.1 shows a pulley, assumed frictionless. Two masses m and M are connected by a light cord passing over the pulley. What is the acceleration of the masses and the tension in the cord?

The mass being accelerated is $M + m$ and so the acceleration a is given by

$$F = (M + m)a$$

$$a = \frac{F}{M + m}$$

But the accelerating force F is $Mg - mg$. Hence

$$a = \frac{(M - m)g}{M + \mathrm{m}}$$

The tension in the cord is the same on both sides. For the side with mass M the resultant force acting on the mass M is

$$Ma = Mg - T$$

and hence

$$T = Mg - Ma$$

For the mass m the resultant force is

$$ma = T - mg$$

and hence

$$T = ma + mg$$

FRICTION

When an object is made to slide across a surface, a force is needed to start the slide. Thus a force can be applied without any motion occurring. This must mean that there is no resultant force and hence the applied force is counterbalanced by an opposite and equal force. This force is called the *frictional force.*

The frictional force is always in such a direction as to oppose the attempted motion. The force is always tangential to the surfaces in contact. The following are some other experimental observations that are reasonably general. The frictional force is independent of the area of the surfaces in contact. The frictional force depends on the materials concerned and its maximum value is directly proportional to the normal reaction between the surfaces in contact. The maximum value, or limiting value as it is generally called, occurs when the applied force is just big enough to start the object sliding.

The ratio of the limiting frictional force F to the normal reaction R is a constant for a particular pair of surfaces and is called the *coefficient of static* or *limiting friction*, μ.

$$\frac{\text{Limiting frictional force } F}{\text{Normal force } N} = \mu$$

or

$$F = \mu N$$

The following are some typical values of the coefficient.

Material	*Static coefficient of friction*
Steel on steel	0.7
Copper on steel	0.5
Glass on glass	0.9

Thus for a steel block sliding on steel there will be a limiting frictional force equal to 0.7 times the normal force. If the block is on a horizontal surface the normal force equals the gravitational force acting on the block, i.e. mg. Thus the limiting frictional force is μmg. Until this force is applied to the block it will not move. If a greater force is applied, the block will accelerate because there is a resultant force acting on the block. The resultant force will be $F - \mu mg$, where F is the applied force. Once sliding occurs the coefficient of fricton tends to be slightly less than the static coefficient. The new coefficient is called the *kinetic* or *dynamic coefficient* of friction.

Material	*Dynamic coefficient of friction*
Steel on steel	0.6
Copper on steel	0.4
Glass on glass	0.4

Example 3 What force has to be applied to a block with a mass of 2 kg to start it moving when it is on a horizontal surface with a coefficient of limiting static friction of 0.6?

$$\text{Normal force} = mg = 2 \times 9.8 \quad \text{kg} \times \text{m/s}^2$$

Hence as

$$\begin{aligned} F &= \mu N \\ &= 0.6 \times 2 \times 9.8 \quad \text{kg} \times \text{m/s}^2 \\ &= 12 \text{ N to two significant figures} \end{aligned}$$

Example 4 What acceleration is produced when a force of 50 N acts on a block of mass 2.5 kg resting on a horizontal surface where the coefficient of dynamic friction is 0.5?

$$\text{Normal force} = 2.5 \times 9.8 \quad \text{kg} \times \text{m/s}^2$$

Hence the frictional force $F = \mu N$

$= 0.5 \times 2.5 \times 9.8 \qquad \text{kg} \times \text{m/s}^2$

$= 12$ N to two significant figures

Hence the resultant force acting on the sliding object is $50 - 12 = 38$ N. Thus the acceleration is given by

Resultant force $= ma$

and so

$$a = \frac{38}{2.5} \qquad \frac{\text{N}}{\text{kg}}$$

$= 15 \text{ m/s}^2$ to two significant figures

MOMENTUM

Momentum is defined as the product of the mass of an object and its velocity. The direction of the momentum is the direction of the velocity.

Momentum $= mv$

The units of momentum are kg m/s.

The significance of momentum is that in any interaction between bodies momentum is always conserved. This is an experimental fact. The principle is known as the *conservation of momentum*. Thus if, as in Figure 5.2, an object with a mass m_1 and velocity v_1 is moving to overtake another object of mass m_2 and velocity v_2 then

Momentum before the collision $= m_1 v_1 + m_2 v_2$

(a) Before the collision

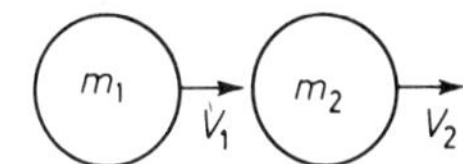

(b) After the collision

Figure 5.2

Both the objects have their velocities in the same direction and thus both the momentums are in the same direction and can be added to give the total momentum at that instant.

Momentum after the collision $= m_1 V_1 + m_2 V_2$

Because momentum is conserved, the momentum before the collision must equal the momentum after the collision.

$$m_1 v_1 + m_2 v_2 = m_1 V_1 + m_2 V_2$$

If any of the momentum in the above equation was in the opposite direction, i.e. the body was moving from right to left, then the momentum would be negative as the direction left to right has been taken as positive.

When a fireman allows water to issue at speed from the jet on a hosepipe, momentum must be conserved. The mass of water acquires a velocity and hence a momentum. This must be balanced by equal momentum in the opposite direction. This momentum is given to the hosepipe. If you have tried to hold a hosepipe from which a high velocity jet of water is emerging, you will have realised that the hosepipe has momentum.

The conservation of momentum equation can be rearranged to give

$$m_1(V_1 - v_1) = -m_2(V_2 - v_2)$$

Thus the change in momentum of one of the objects is equal in size to the change of momentum of the other object. When the two objects collide they each change in velocity. This change in velocity takes place in some time, say t. This time must be the same for both objects, after all they must both have the same time in contact during the collision, so

$$\frac{m_1(V_1 - v_1)}{t} = -\frac{m_2(V_2 - v_2)}{t}$$

The rate of change of momentum is thus the same for both objects. But when the velocity of an object changes it accelerates, acceleration being the rate of change of velocity. Thus, for uniform acceleration,

$$a_1 = \frac{V_1 - v_1}{t}$$

and

$$a_2 = \frac{V_2 - v_2}{t}$$

Hence

$$m_1 a_1 = -m_2 a_2$$

But the product of mass and acceleration is force. Thus the force experienced by one object during the collision must be equal and opposite to the force experienced by the other object. This is Newton's third law. The evidence for the third law can thus be based on the experimental fact that momentum is conserved.

Hence, with the hosepipe mentioned earlier, the force which results in the water being accelerated and emerging from the hosepipe nozzle with velocity must be balanced by an opposite and equal force acting on the nozzle of the hosepipe.

The force experienced by the hosepipe nozzle or indeed any object when its momentum changes is equal to the rate of change of momentum.

Force = rate of change of momentum

Example 4 A truck with a mass of 40 kg and moving with a velocity of 4 m/s collides with a truck of mass 60 kg moving in the same straight line but in the opposite direction with a velocity of 2 m/s. When the trucks collide they lock together. With what velocity will they move after the collision?

$$\text{Momentum before the collision} = 40 \times 4 - 60 \times 2 \quad \text{kg} \times \text{m/s}$$
$$= 40 \text{ kg m/s}$$

The 60 kg truck is considered to have a negative momentum because it is moving in the opposite direction to the 40 kg truck.

$$\text{Momentum after the collision} = (40 + 60)V$$

Hence, as momentum is conserved

$$(40 + 60)V = 40$$
$$V = 0.4 \text{ m/s}$$

The velocity is positive thus indicating that it is in the direction of the velocity of the 40 kg truck as this was taken as being positive.

Example 5 A bullet of mass 2 g emerges from the barrel of a gun with a velocity of 300 m/s. What is the recoil momentum of the gun?

$$\text{Momentum of the bullet} = 2 \times 10^{-3} \times 300 \quad \text{kg} \times \text{m/s}$$
$$= 0.6 \text{ kg m/s}$$

The momentum acquired by the gun must be –0.6 kg m/s, the minus sign indicating that it is in the opposite direction to that of the bullet.

Example 6 Calculate the average recoil force experienced by a machine gun firing 120 shots per minute, each bullet having a mass of 2 g and a muzzle velocity of 300 m/s.

$$\text{Momentum of 1 bullet} = 2 \times 10^{-3} \times 300 \quad \text{kg m/s}$$
$$= 0.6 \text{ kg m/s}$$
$$\text{Momentum of 120 bullets} = 120 \times 0.6$$
$$= 72 \text{ kg m/s}$$

This is the momentum given to the bullets in 60 seconds. In 1 s the momentum given to the bullets must be 1.2 kg m/s. This is the rate at which bullets receive momentum and must also be the rate at which the gun receives momentum. The force experienced by the gun will equal the rate of change of momentum.

$$\text{Force} = \text{rate of change of momentum}$$
$$= 1.2 \text{ kg m/s/s}$$
$$= 1.2 \text{ N}$$

VELOCITY

When an object is moving with a constant velocity there is no resultant force acting on the object. For all other conditions, there is a force acting on the body. It is thus important to recognise what is meant by a constant velocity. A *constant velocity* is defined as the coverage of equal distances in the same straight line in equal intervals of time, no matter how small a time interval is considered. An important part of that definition is the term 'same straight line'. Because it is important to know not only the rate at which distance is covered but also the direction, velocity being a vector quantity.

If equal distances are covered in equal intervals of time, no matter how small a time interval is considered, but the direction might vary, the speed is said to be constant but not the velocity. Speed is a scalar quantity.

An object has an acceleration when its velocity changes. It can, however, have an acceleration even if its speed is constant, since there could be a change in direction. An object moving at constant speed round a circular path has an acceleration and, therefore, a force is acting on it. If there was no force the object would not move in a circular path but in a straight line.

A person walks along the aisle in a train with a velocity of 2 m/s relative to the train. If the train is moving at 60 m/s relative to the track then the velocity of person relative to the track is 62 m/s if he is walking towards the front of the train and 58 m/s if he is walking towards the rear of the train. The direction of the velocity has to be taken into account in determining relative velocity.

In the above instance, the velocities were in the same straight line. Where they are not in the same straight line the normal methods associated with vectors, e.g. those used with forces in Chapter 1, need to be used. Thus, if a tool traverses a lathe bed at 1.5 mm/s relative to the

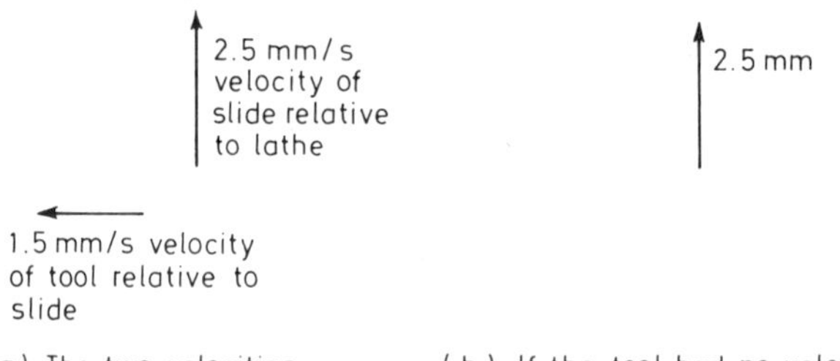

(a) The two velocities

(b) If the tool had no velocity relative to the slide it would move 2.5 mm in this direction in 1 s

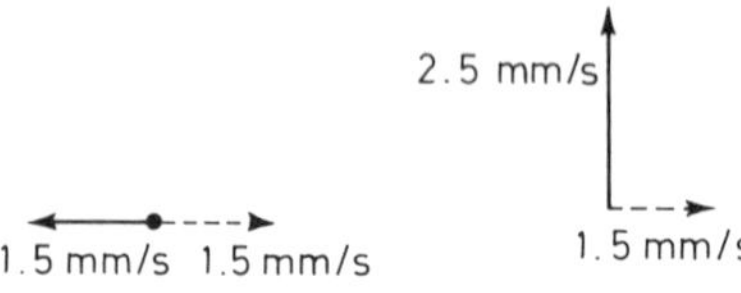

(c) We can simplify the problem by adding an opposite and equal velocity to the tool to bring it to rest relative to the slide. We must also add the same velocity to the slide.

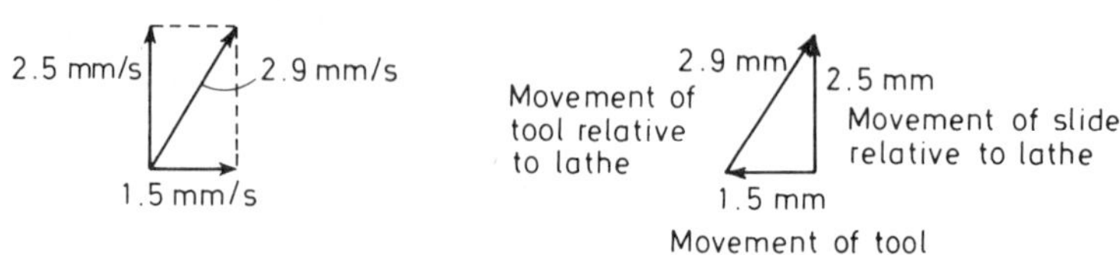

(d) The velocity of the tool is the same as the modified velocity of the slide, i.e. 2.9 mm/s

(e) The triangle effectively represents the velocities concerned

Figure 5.3

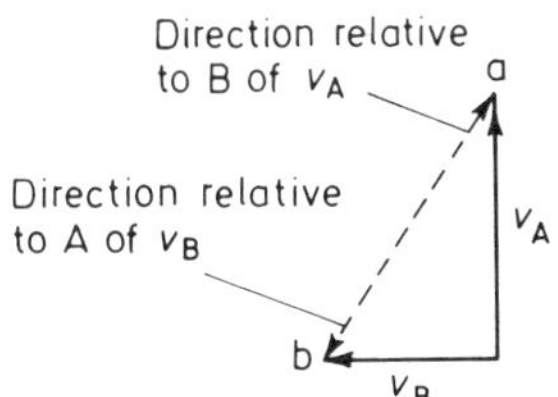

Figure 5.4

slide and the slide itself is moving at 2.5 mm/s relative to th a direction at right angles to the tool velocity a velocity di... be used to give the velocity of the tool relative to the lathe. The tool moves in 1 s a distance of 1.5 mm in one direction and 2.5 mm in a right angle direction (Figure 5.3). The actual distance moved relative to the lathe is 2.9 mm (to two significant figures) in the direction indicated in the diagram.

The method used to obtain the relative velocity above can be summarised as follows: Represent the magnitude and direction of the velocities by lines. Draw each of the lines from the same point. The line which completes the triangle gives the relative velocity (Figure 5.4).

WORK AND ENERGY

When a force causes a body to move it is said to do *work*. Work is defined as the product of the force and the distance travelled by the body in the direction of the force.

$$\text{Work} = \text{force} \times \text{distance}$$

The unit of work is the N m or joule (J).

Thus if an object is pulled through a distance of 0.10 m against the action of a spring which exerts an average force of 100 N along the line of movement, then the work done is $100 \times 0.10 = 10$ J.

If an object with a mass of 2 kg is lifted through a vertical distance of 0.8 m, work is done because the object is being lifted against the gravitational force which acts on the object.

$$\begin{aligned}\text{Gravitational force} &= 2 \times 9.8 \\ &= 19.6 \text{ N}\end{aligned}$$

Hence

$$\begin{aligned}\text{Work} &= 19.6 \times 0.8 \qquad \text{N} \times \text{m} \\ &= 15.7 \text{ J to three significant figures}\end{aligned}$$

When the object has been lifted to the height of 0.8 m, it is said to have gained *potential energy*. If the object was released, it would fall to the ground and lose its height of 0.8 m. With the spring and the object being pulled against the force of the spring, when the object had been pulled 0.10 m it is said to have gained potential energy. When released, it would be pulled back to its initial position and lose its 0.10 m distance for which work was expended. In both the case of the object being lifted against the gravitational force and the object being pulled against the action of the spring force, work has been expended to move the objects to the required positions. When the objects are released, they move back to their original positions. We talk of work being the transfer of energy, the objects in these cases gaining energy. They gain the energy and move to a new position. The energy associated with these new positions is called *potential energy*.

The gravitational force acting on a mass m is mg, where g is the acceleration due to gravity. When such an object is lifted through a vertical height of h the work done is $mg \times h$. This is the energy given to the object and held by it by virtue of its position. This is the potential energy.

$$\text{Potential energy} = mgh$$

The unit of potential energy is the joule.

When an object falls, it loses potential energy. It does, however, accelerate as it falls and so increases its velocity. The object, when falling, is moving under the action of a force, the gravitational force. Instead of gaining height by moving against the force it is moving with the force and gaining velocity. We say it is gaining kinetic energy. *Kinetic energy* is the energy a body possesses by virtue of its motion. When the falling object has lost all its potential energy and gained kinetic energy it will have fallen through a distance h.

Potential energy lost $= mgh$

Force acting on the falling object $= mg$

Distance through which it falls $= h$

Thus the potential energy lost is the work involved when the object falls. It would thus seem reasonable to say that the potential energy lost must equal the kinetic energy gained.

Kinetic energy gained $= mgh$

This can be converted into an expression involving the velocity attained as a result of the fall.

$$v^2 = u^2 + 2as$$

The above equation applies for an object starting with a velocity u and being accelerated at a constant acceleration a over a distance s, the velocity after that distance being v. For the falling object the acceleration is g and as it falls from rest $u = 0$. Thus for a fall through a distance h

$$v^2 = 2gh$$

Hence

$$\text{Kinetic energy gained} = mg\frac{v^2}{2g}$$

$$= \tfrac{1}{2}mv^2$$

Work is the process by which energy is transferred to an object, which gains energy as a result of the transfer. The gain could be in the form of potential energy or kinetic energy. The rate at which the energy is transferred is called the *power.*

Power = rate of transfer of energy

The unit of power is the joule/s or watt.

Example 7 The locomotive of a train exerts a constant force of 120 kN on a train while pulling it at 40 km/h along a level track. What work will the locomotive do in 15 minutes?

In 15 minutes the train will cover a distance of 10 km.

Thus the work done $= 120 \times 10^3 \times 10 \times 10^3 \quad \text{N} \times \text{m}$

$= 1.2$ GJ

Example 8 Calculate the kinetic energy of a bullet of mass 2 g travelling at 300 m/s.

$$\text{Kinetic energy} = \tfrac{1}{2}mv^2$$
$$= \tfrac{1}{2} \times 2 \times 10^{-3} \times 300^2 \qquad \text{kg} \times (\text{m/s})^2$$
$$= 90\ \text{J}$$

Example 9 A 3 kg block is lifted vertically through a height of 2m. How much potential energy does the block gain?

$$\text{Potential energy} = mgh$$
$$= 3 \times 9.8 \times 2 \qquad \text{kg} \times \text{m/s}^2 \times \text{m}$$
$$= 49.8\ \text{J}$$

Example 10 If the block in Example 9 had been lifted with a constant velocity of 3 m/s through the height of 2 m, what would have been the power?

This requires a calculation of the time taken for the energy transfer to occur.

$$\text{Time taken} = \frac{\text{distance covered}}{\text{velocity}}$$
$$= \frac{2}{3} \qquad \frac{\text{m}}{\text{m/s}}$$
$$= \tfrac{2}{3}\ \text{s}$$

In this time the work done was 49.8 J, thus the power is given by

$$\text{Power} = \frac{49.8}{\frac{2}{3}} \qquad \frac{\text{J}}{\text{s}}$$
$$= 74.7\ \text{W}$$

PROBLEMS

1. State Newton's laws of motion.

2. Calculate the forces needed for the following: (a) an object of mass 200 g is to be accelerated at 2 m/s^2, (b) a truck of mass 40 kg is to be accelerated at 0.3 m/s^2, (c) a bullet of mass 2 g is to be accelerated to 300 m/s in 0.1 s.

3. A constant horizontal force of 50 N acts on a body on a smooth horizontal plane. The body starts from rest and is observed to move 75 m in 5 s. What is the mass of the body?

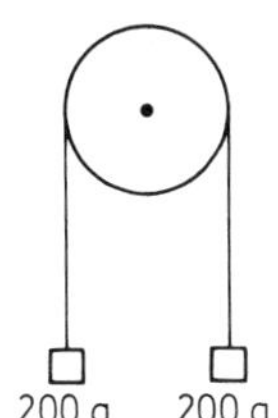

Figure 5.5

4. Two 200 g blocks hang at the ends of a light flexible rope which passes over a small frictionless pulley (Figure 5.5). What is the tension in the rope?

5. What is the tension in the rope passing over a small frictionless pulley if at one end of the rope there is a mass of 200 g and at the other end a mass of 100 g? Assume the rope to be light and flexible, the arrangement being of the form shown in Figure 5.6. What is the acceleration?

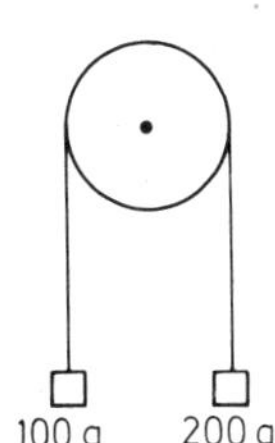

Figure 5.6

6. With what acceleration will the blocks move for the arrangement described in Problem 5 if the 200 g mass is replaced by 300 g?

7. A man with a mass of 80 kg stands on a platform which itself has a mass of 40 kg. The man pulls the platform, and himself, upwards with an acceleration of $0.6\ m/s^2$ by means of pulling on a rope which runs over a pulley above his head. With what force does he need to pull on the rope?

8. What is the momentum of a bullet of mass 3 g when moving with a velocity of 340 m/s?

9. What is the momentum of a 10 000 kg truck moving with a velocity of 60 km/h?

10. A car of mass 1500 kg moving at 50 km/h collides head on with a truck of mass 4000 kg coming in the opposite direction with a velocity of 30 km/h. If the two become locked together after the collision what will be their velocity?

11. A gun of mass 150 000 kg fires a shell of mass 1200 kg, giving it a muzzle velocity of 800 m/s. What is the initial recoil velocity of the gun?

12. Explain the term momentum and what is meant by the conservation of momentum.

13. What force is needed to start a block of steel of mass 5 kg moving along a horizontal steel surface if the coefficient of limiting friction is 0.6?

14. What is the acceleration produced when a force of 40 N acts on a block of steel of mass 2 kg resting on a horizontal surface where the coefficient of friction is 0.5?

15. A block of steel rests on an incline which is at an angle of 30° to the horizontal. The block has a mass of 5 kg. (a) What is the gravitational force acting on the block? (b) What is the reaction force at the surface? (c) What is the component of the gravitational force acting down the slope? (d) If the block just starts to slip under its own weight what must be the coefficient of friction?

16. An aircraft flies due west with a velocity of 500 km/h. Another aircraft is flying due north with a velocity of 400 km/h. What is the velocity of the second aircraft relative to the first? What is the relative velocity of the first aircraft relative to the second?

17. The pilot of an aircraft wishes to fly due north. With a wind of 70 km/h blowing from the west and an aircraft with a flying speed of 250 km/h in still air, in what direction should the pilot aim to fly?

18. A block of metal of mass 20 kg is hoisted vertically through a distance of 5 m. What is the work done in lifting the block?

19. A load with a mass of 2000 kg is hauled up an incline of 1 in 100. The frictional resistance opposing the motion up the plane is constant at 300 N. What energy is needed to get the load a distance of 4 m up the slope?

20. An object of mass 200 g falls from rest a height of 2000 mm onto the floor. With what velocity will it hit the floor? What work is needed to lift the object back up off the floor to its initial height?

21. A cutting tool operates against a constant resistive force of 2000 N.

If the tool moves through a distance of 150 mm in 6 s what is the pow used?

22. What power is involved in a car with a weight of 8000 N travelling at a constant velocity of 60 km/h up an incline of 1 in 50?

23. A car engine develops a power of 6 kW when the car, of mass 1600 kg, is climbing a hill with a gradient of 1 in 60. Calculate the size of the resistive forces.

24. In a stamping operation a force of 20 kN is used to compress an assembly by 1 mm. What is the energy required for the operation?

25. A drop hammer of mass 500 kg falls from rest through a vertical height of 5 m onto a forging. With what kinetic energy will the hammer hit the forging?

6 Angular motion

ANGULAR VELOCITY AND ACCELERATION

When an object moves in a straight line we are able to refer to the *linear velocity* of that object as the rate at which distance is covered in that straight line.

$$\text{Linear velocity } v = \frac{\text{distance covered in a straight line}}{\text{time taken for that distance}}$$

If the object covered equal distances in the same straight line in equal intervals of time, no matter how small a time interval we consider, then the linear velocity is said to be constant or uniform.

With linear velocity we are concerned with distance measured in metres or multiples of metres. When we deal with angular motion we become concerned with angles rather than distances. When an object rotates about a point it sweeps out an arc. The arc length divided by the radius of the rotation defines the angle in *radians.*

$$\text{Angle in radians } \theta = \frac{\text{arc length}}{\text{radius}}$$

An angle in radian measure is a ratio of two lengths and thus is without units. The abbreviation for radian is rad.

One complete rotation involves an arc length equal to the circumference of a circle, i.e. $2\pi r$. Thus with r being the radius of the rotation

$$\theta = \frac{2\pi r}{r} = 2\pi \text{ radians}$$

Thus, one complete revolution is a rotation through 2π radians. A rotation through π radians is thus a rotation through half the complete circle, i.e. 180°.

When we have rotational motion about an axis we are able to refer to the *angular velocity* as the rate at which angle is covered or swept out.

$$\text{Angular velocity } \omega = \frac{\text{angle rotated through}}{\text{time taken for that rotation}}$$

The unit of angular velocity is rad/s.

If the object rotates through equal angles in equal intervals of time, no matter how small a time interval we consider, then the angular velocity is said to be constant or uniform.

The angular velocity is related to the *frequency* or the number of revolutions made per second. Thus if f revolutions are made per second then in one second the angle rotated through will be $2\pi f$. But the angle rotated through per second is the angular velocity. Thus

$$\text{Angular velocity } \omega = 2\pi f$$

Linear acceleration is the rate at which linear velocity changes.

$$\text{Linear acceleration } a = \frac{\text{change in linear velocity}}{\text{time taken for the change}}$$

If the linear velocity changes by equal amounts in equal intervals of time, no matter how small a time interval we consider, then the linear acceleration is said to be constant or uniform.

Angular acceleration is the rate at which angular velocity changes.

$$\text{Angular acceleration } \alpha = \frac{\text{change in angular velocity}}{\text{time taken for the change}}$$

The unit of angular acceleration is rad/s^2.

If the angular acceleration changes by equal amounts in equal intervals of time, no matter how small a time interval we consider, then the angular acceleration is said to be constant or uniform.

Example 1 A flywheel rotates through 6 complete revolutions in 3 s. What is the angular velocity over that time interval?

$$\text{Angle rotated through in 3 s} = 6 \times 2\pi \quad \text{rad}$$

Hence

$$\text{Angular velocity} = \frac{6 \times 2\pi}{3} \quad \frac{\text{rad}}{\text{s}}$$

$$= 4\pi \text{ rad/s}$$

$$= 13 \text{ rad/s to two significant figures}$$

Example 2 The angular velocity of a flywheel changes from 8 rad/s to 12 rad/s in 2 s. What is the angular acceleration if it is assumed to be uniform over the time interval concerned?

$$\text{Angular acceleration} = \frac{\text{change in angular velocity}}{\text{time taken for the change}}$$

$$= \frac{12 - 8}{2} \quad \frac{\text{rad/s}}{\text{s}}$$

$$= 2 \text{ rad/s}^2$$

RATES OF CHANGE

If a quantity x changes from x_1 at time t_1 to x_2 at time t_2 at a uniform rate, then the rate of change of x with t is

$$\text{Rate of change} = \frac{x_2 - x_1}{t_2 - t_1}$$

The change in the value of x can be represented as Δx, similarly the change in t can be represented as Δt. Thus,

$$\Delta x = x_2 - x_1$$

$$\Delta t = t_2 - t_1$$

Hence

$$\text{Rate of change} = \frac{\Delta x}{\Delta t}$$

If x is not changing uniformly with time the value given by the above equation will be the average rate of change over the interval concerned. If we want the instantaneous rate of change we need to consider vanishingly small intervals. We write such vanishingly small intervals as

$$\text{Rate of change} = \frac{dx}{dt}$$

Linear velocity is the rate of change of displacement with time. Thus for the instantaneous velocity where x represents displacement and t time we have

$$\text{Linear velocity} = \frac{dx}{dt}$$

Linear acceleration is the rate of change of velocity with time. Thus for the instantaneous acceleration where v represents velocity and t time we have

$$\text{Linear acceleration} = \frac{dv}{dt}$$

Angular velocity is the rate at which angle is covered with time. Thus for the instantaneous angular velocity where θ represents angle and t time we have

$$\text{Angular velocity} = \frac{d\theta}{dt}$$

Angular acceleration is the rate at which angular velocity varies with time. Thus for the instantaneous angular acceleration where ω represents angular velocity and t time we have

$$\text{Angular acceleration} = \frac{d\omega}{dt}$$

These rates of change represent the slopes of graphs at a specific point. Thus for a graph of displacement against time the rate of change, or instantaneous velocity, is the slope of the graph at a particular time. The quantity dx/dt is the slope. Similarly for angular velocity $d\theta/dt$, this represents the slope of the graph of θ against time at a particular time.

The methods of differential calculus can be used to evaluate values of rates of change from equations relating the quantities concerned. An example of this is given for simple harmonic motion in Chapter 7.

ROTATION WITH CONSTANT ANGULAR ACCELERATION

With constant angular acceleration the angular velocity is changing by the same amount in equal intervals of time. Thus if at the beginning of some time interval the angular velocity was ω_0 and at the end of that time interval t the angular velocity was ω as a result of a constant angular acceleration α, then

$$\alpha = \frac{\text{change in angular velocity}}{\text{time taken for the change}}$$

$$a = \frac{\omega - \omega_0}{t}$$

Hence

$$\omega = \omega_0 + \alpha t \tag{6.1}$$

This equation can be compared with the equation for linear acceleration a from a velocity u to v in time t.

$$v = u + at$$

When the angular velocity changes from ω_0 to ω in a time t with a constant angular acceleration α the average angular velocity over that time interval will be

$$\text{Average angular velocity} = \frac{\omega_0 + \omega}{2}$$

If in the time t the angle covered was θ then

$$\text{Average angular velocity} = \frac{\theta}{t}$$

thus

$$\frac{\theta}{t} = \frac{\omega_0 + \omega}{2}$$

$$\theta = \frac{\omega_0 t}{2} + \frac{\omega t}{2}$$

But the earlier Equation (6.1) gave ω as $\omega_0 + \alpha t$, thus

$$\theta = \frac{\omega_0 t}{2} + \frac{t}{2}(\omega_0 + \alpha t)$$

$$\theta = \omega_0 t + \tfrac{1}{2}\alpha t^2 \qquad (6.2)$$

This equation can be compared with the equation for distance s covered with a linear acceleration

$$s = ut + \tfrac{1}{2}at^2$$

If t is eliminated from the two simultaneous equations (6.1) and (6.2) we obtain

$$\theta = \omega_0 \frac{(\omega - \omega_0)}{\alpha} + \tfrac{1}{2}\alpha \frac{(\omega - \omega_0)^2}{\alpha^2}$$

$$2\alpha\theta = 2\omega_0\omega - 2\omega_0^2 + \omega^2 - 2\omega\omega_0 + {\omega_0}^2$$

or

$$\omega^2 = {\omega_0}^2 + 2\alpha\theta \qquad (6.3)$$

This equation can be compared with the equation for the distance s covered with linear acceleration a.

$$v^2 = u^2 + 2as$$

The equations (6.1), (6.2) and (6.3) are called the *equations of angular motion.* They all assume that the angular acceleration is constant.

Example 3 A body is rotating with an angular velocity of 4 rad/s, when it is accelerated with an angular acceleration of 2 rad/s^2. What will be the angular velocity after 3 s?

$$\omega = \omega_0 + \alpha t$$

$$\omega = 4 + 2 \times 3$$

$$= 10 \text{ rad/s}$$

Example 4 A flywheel takes 5 s to rotate through 200 rad starting from rest. If it is accelerating with a constant angular acceleration what is its value?

$$\theta = \omega_0 t + \tfrac{1}{2}\alpha t^2$$

$$200 = 0 + \tfrac{1}{2}\alpha \times 5^2$$

$$\alpha = 8\ \mathrm{rad/s^2}$$

Example 5 A wheel starts from rest and accelerates uniformly with an angular acceleration of 4 rad/s². What will be its angular velocity after 4 s and the total angle rotated in that time?

$$\omega = \omega_0 + \alpha t$$

$$\omega = 0 + 4 \times 4$$

$$\omega = 16\ \mathrm{rad/s}$$

For the total angle calculation,

$$\theta = \omega_0 t + \tfrac{1}{2}\alpha t^2$$

$$\theta = 0 + \tfrac{1}{2} \times 4 \times 4^2$$

$$\theta = 32\ \mathrm{rad}$$

RELATIONSHIP BETWEEN LINEAR AND ANGULAR MOTION

A point moving round the arc of a circle of radius r moves through an arc of length s which subtends an angle θ at the centre of the circular path.

$$s = r\theta$$

If the point is moving with a constant angular velocity ω then in time the angle θ covered will be given by

$$\omega = \frac{\theta}{t}$$

and so

$$\theta = \omega t$$

Thus

$$s = r \times \omega t$$

and

$$\frac{s}{t} = r\omega$$

But, s/t is the speed, and if very small time intervals are considered then it is the velocity v at the point concerned.
Hence

$$v = r\omega$$

The linear velocity is the angular velocity multiplied by the radius of the angular path.

If the point has a linear velocity of u initially then its angular velocity ω_0 at that point in its angular path will be given by

$$u = r\omega_0$$

If at some later point in its path after a uniform angular acceleration the linear velocity is v then the angular velocity ω will be given by

$$v = r\omega$$

The angular acceleration α is given by

$$\alpha = \frac{\omega - \omega_0}{t}$$

and so

$$r\alpha = \frac{r\omega - r\omega_0}{t}$$

$$r\alpha = \frac{v - u}{t}$$

But the linear acceleration a is given by

$$a = \frac{v - u}{t}$$

Thus

$$a = r\alpha$$

The linear acceleration is the angular acceleration multiplied by the radius of the angular path.

Example 6 A wheel has a radius of 200 mm. Calculate the linear velocity of a point on the rim when the wheel is rotating with an angular velocity of 5 rad/s.

$$\begin{aligned} v &= r\omega \\ &= 200 \times 5 \qquad \text{mm} \times \text{rad/s} \\ &= 1000 \text{ mm/s} \end{aligned}$$

Example 7 The linear speed of a belt that passes round a pulley wheel is 20 m/s. If there is no slip, what will be the angular velocity of the 150 mm radius pulley? What will be the speed of the pulley in revolutions per minute?

$$v = r\omega$$

Hence

$$\begin{aligned} \omega &= \frac{20}{0.15} \qquad \frac{\text{m/s}}{\text{m}} \\ &= 130 \text{ rad/s to two significant figures} \end{aligned}$$

The angular velocity $\omega = 2\pi f$, where f is the frequency of rotation. Thus

$$\begin{aligned} f &= \frac{130}{2\pi} \qquad \text{s}^{-1} \\ &= 21 \text{ s}^{-1} \text{ to two significant figures} \\ &= 1260 \text{ min}^{-1} \end{aligned}$$

Example 8 A flywheel accelerates from 20 to 40 revolutions per minute in 8 s. If the radius of the wheel is 2.5 m calculate (a) the angular acceleration and (b) the linear acceleration of a point on the rim of the wheel.

(a) The angular velocity changes from $2\pi \times 20\ \text{min}^{-1}$ to $2\pi \times 40\ \text{min}^{-1}$ in 8 s. This is a change from $2\pi \times 20\ \text{s}^{-1}/60$ to $2\pi \times 40\ \text{s}^{-1}$, thus the angular acceleration is

$$\text{Angular acceleration} = \frac{2\pi(40-20)}{60 \times 8} \qquad \frac{\text{s}^{-1}}{\text{s}}$$

$$= 0.52\ \text{s}^{-2} \text{ to two significant figures}$$

(b)

$$a = r\alpha$$

$$= 2.5 \times 0.52 \qquad \text{m} \times \text{s}^{-2}$$

$$= 1.3\ \text{m/s}^2$$

TORQUE AND ANGULAR ACCELERATION

Consider a simple case involving rotation – a small object of mass m at the end of a pivoted rod (Figure 6.1). A force F applied to the mass and at right angles to the rod will cause the mass to move in a circular path of radius r. The torque causing the rotation will be Fr.

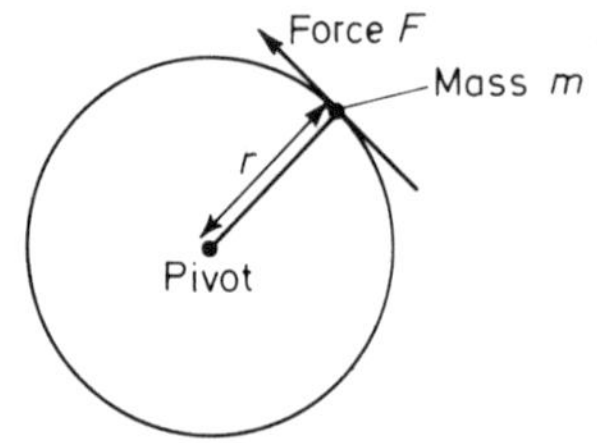

Figure 6.1

$$\text{Torque } T = Fr$$

But the force F acting on the mass must cause the mass to accelerate with a linear acceleration a.

$$F = ma$$

Thus

$$T = mar$$

The acceleration will be in the direction of the force. The linear acceleration a will be related to the angular acceleration α by

$$a = r\alpha$$

Hence

$$T = mr^2\alpha$$

The angular acceleration is thus proportional to the torque. The quantity mr^2 determines the angular acceleration produced for a given torque. The bigger the value of mr^2 the smaller the angular acceleration for a given torque. The quantity mr^2 thus represents the difficulty in producing a given angular acceleration. It is thus like the mass term in $F = ma$ and represents the inertia of the system. The bigger the mass m the more difficult it is to produce a linear acceleration for a given force.

The quantity mr^2 is called the *moment of inertia I*.

$$T = I\alpha$$

The unit of moment of inertia is N m/s or kg m^2.

For a small mass at the end of a pivoted arm of radius r

$$I = mr^2$$

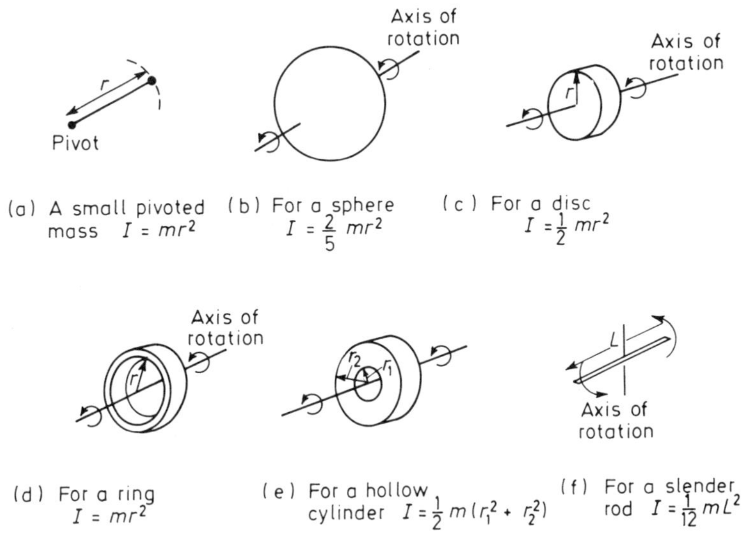

(a) A small pivoted mass $I = mr^2$ (b) For a sphere $I = \frac{2}{5}mr^2$ (c) For a disc $I = \frac{1}{2}mr^2$

(d) For a ring $I = mr^2$ (e) For a hollow cylinder $I = \frac{1}{2}m(r_1^2 + r_2^2)$ (f) For a slender rod $I = \frac{1}{12}mL^2$

Figure 6.2 Moments of inertia

If the rotating object is more complicated than a point mass the moment of inertia about some pivot point is the sum of all the mr^2 terms due to each small particle of mass that makes up the body. Figure 6.2 shows moments of inertia for some common bodies.

Whatever the form of a body and however its mass is distributed it is always possible to consider some radial distance from the pivot point at which all the mass can be considered to act. Thus, for example, for the disc shown in Figure 6.2(c), the moment of inertia is given by

$$I = ½mr^2$$

If all the mass of that body had been considered to have been at a distance k from the pivot point then we would have had

$$I = mk^2$$

as in the case of the point mass at the end of the pivoted arm considered earlier in this section. This would thus seem to indicate that

$$k^2 = ½r^2$$

$$k = \sqrt{0.5} \times r$$

$$k = 0.707r \text{ to three significant figures}$$

Thus for the disc we can consider all the mass to be located at a point $0.707r$ from the pivot point. This distance k is called the *radius of gyration.*

For the objects shown in Figure 6.2 the radii of gyration are:

sphere	$k = \frac{2}{5}r$
disc	$k = \sqrt{½}r$
ring	$k = r$
hollow cylinder	$k = \sqrt{\{½(r_1{}^2 + r_2{}^2)\}}$
slender rod	$k = \sqrt{½}r$

Example 9 What torque is required to cause a pulley to rotate at 5 revolutions per second in 15 s starting from rest? The pulley has a moment of inertia of 200 kg m^2 and the acceleration may be assumed to be uniform.

$$\omega = 2\pi f$$

$$\omega = 2\pi \times 5 \qquad \text{s}^{-1}$$

The angular acceleration can be calculated from

$$\omega = \omega_0 + \alpha t$$

$$10\pi = 0 + 15\alpha$$

$$\alpha = \frac{10\pi}{15} \qquad \text{s}^{-2}$$

Hence the torque is given by

$$T = I\alpha$$

$$= 200 \times \frac{10\pi}{15} \qquad \text{kg m}^2\ \text{s}^{-2}$$

$$= 420 \text{ N m to two significant figures}$$

Example 10 A flywheel has a moment of inertia of 10 kg m^2. What will be the angular acceleration produced by a torque of 5 N m?

$$T = I\alpha$$

Hence

$$\alpha = \frac{5}{10} \qquad \frac{\text{N m}}{\text{kg m}^2} = \frac{\text{kg m}}{\text{s}^2} \times \frac{\text{m}}{\text{kg m}^2}$$

$$\alpha = 0.5 \text{ rad/s}^2$$

Example 11 If for the flywheel in the above question there had been bearing friction equivalent to a torque of 2 N m what would the angular acceleration have been?

Accelerating torque = 5 – 2 = 3 N m

Hence

$$\alpha = \frac{3}{10}$$

$$= 0.3 \text{ rad/s}^2$$

Example 12 An electric motor has a rotor of mass 200 kg and radius of gyration of 140 mm. Calculate the torque required to accelerate the rotor at 0.5 s^{-2}.

$$I = mk^2$$

$$= 200 \times 0.14^2 \qquad \text{kg m}^2$$

Hence

$$T = I\alpha$$
$$= 200 \times 0.14^2 \times 0.5 \quad \text{kg m}^2\ \text{s}^{-2}$$
$$= 2.0 \text{ N m to two significant figures}$$

WORK DONE BY A TORQUE

When a force is used to rotate an object then work is done, since a force has moved an object through a distance. Thus in Figure 6.3 the force F is used to move an object so that the end of it moves from A to B while it sweeps out an angle θ. The distance moved is the arc AB.

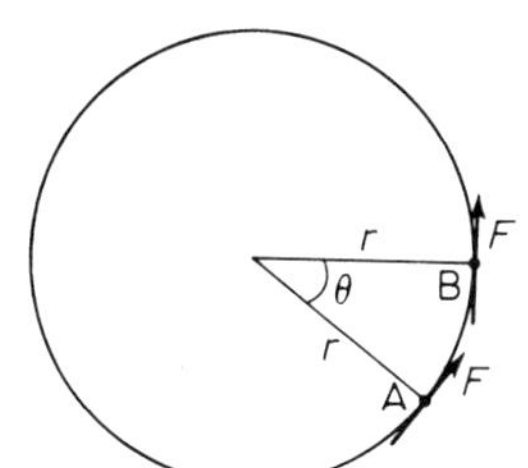

Figure 6.3

$$\text{arc AB} = r\theta$$

Thus

$$\text{Work done} = \text{force} \times \text{distance}$$
$$= F \times \text{arc AB}$$
$$= Fr\theta$$

But the torque T is Fr, thus

$$\text{Work done} = T\theta$$

If the object rotates at a constant f revolutions per second under the action of this torque then in one second the angle turned through will be

$$\theta = 2\pi f$$

Hence the work done in one second will be $T \times 2\pi f$. But the work done per second is the power. Hence

$$\text{Power} = T \times 2\pi f$$

If the torque T is used to keep an object rotating with a constant angular velocity ω then the power is

$$\text{Power} = T\omega$$

since $\omega = 2\pi f$.

Example 13 A constant torque of 50 N m is used to keep a flywheel rotating at a constant 5 revolutions per second. What is the power input to the flywheel?

$$\text{Power} = T \times 2\pi f$$
$$= 300 \times 2\pi \times 5 \quad \text{N m} \times \text{s}^{-1}$$
$$= 9400 \text{ W, to two significant figures}$$
$$= 9.4 \text{ kW}$$

KINETIC ENERGY OF ROTATION

When an object with a mass m has a linear velocity v then it is said to have a kinetic energy of $\frac{1}{2}mv^2$ (see Chapter 5). There is kinetic energy associated with the object because it has a linear velocity. A kinetic energy can also be associated with rotating bodies when they have angular velocities.

Consider the work done in accelerating a flywheel from rest to an angular velocity of ω with a constant angular acceleration of α.

$$\omega^2 = \omega_0{}^2 + 2\alpha\theta$$

See earlier this chapter for the derivation of the above equation.

Thus in this case

$$\omega^2 = 2\alpha\theta$$

Hence the torque required for this acceleration is given by

$$T = I\alpha$$

where I is the moment of inertia about the axis of the flywheel. Thus

$$T = I\frac{\omega^2}{2\theta}$$

The work done by a torque T in rotating the flywheel through an angle of θ is

$$\text{Work done} = T\theta$$

Hence

$$\text{Work done} = I\frac{\omega^2}{2\theta}\,\theta$$
$$= \tfrac{1}{2}I\omega^2$$

The work done in accelerating the flywheel results in a gain in angular velocity of that flywheel. This is a gain in energy of the flywheel. Work is the transfer of energy and thus the energy transferred to the flywheel must be $\tfrac{1}{2}I\omega^2$. The energy is kinetic energy.

$$\text{Kinetic energy of rotation} = \tfrac{1}{2}I\omega^2$$

UNIT: joule (J)

If the flywheel, or indeed any rotating system, had an initial angular velocity of ω_0 and was accelerated to an angular velocity of ω then the work done in producing this change would be the difference in the kinetic energy the wheel had at the end of the acceleration and the kinetic energy it had initially.

$$\text{Work done} = T\theta = \tfrac{1}{2}I\omega^2 - \tfrac{1}{2}I\omega_0{}^2$$

Example 14 A flywheel has a mass of 50 kg and a radius of gyration of 400 mm. Calculate (a) the uniform torque needed to increase the rate of revolution from 200 to 250 revolutions per minute in 20 s and (b) the gain in kinetic energy as a result.

(a) The moment of inertia of the flywheel $I = mk^2$

$$I = 50 \times \frac{400}{1000}{}^2 \quad \text{kg m}^2$$
$$= 8 \text{ kg m}^2$$

The initial angular velocity was $2\pi \times 200/60 \text{ s}^{-1}$ and the final angular velocity $2\pi \times 250/60 \text{ s}^{-1}$. The angular acceleration is thus

$$\alpha = \frac{2\pi \times 250/60 - 2\pi \times 200/60}{20} \quad \frac{s^{-1}}{s}$$

$$= \frac{2\pi \times 50}{60 \times 20} \quad s^{-2}$$

Hence the torque T required to produce this acceleration is given by

$$T = I\alpha$$

$$= 8 \times \frac{2\pi \times 50}{60 \times 20} \quad kg\ m^2\ s^{-2}$$

$= 2.1$ N m to two significant figures

(b) The gain in kinetic energy is given by

$$\text{Gain} = \tfrac{1}{2}I\omega^2 - \tfrac{1}{2}I\omega_0{}^2$$

$$= \tfrac{1}{2}I(\omega^2 - \omega_0{}^2)$$

$$= \tfrac{1}{2} \times 8 \times \left(\frac{2\pi \times 250}{60} - \frac{2\pi \times 200}{60}\right)^2 \quad kg\ m^2 \times s^{-2}$$

$$= \frac{8 \times 4\pi^2}{2 \times 60^2}(250^2 - 200^2)$$

$= 990$ J to two significant figures

Example 15 A flywheel has a mass of 300 kg and a radius of gyration of 1.2 m. Calculate the energy stored in the flywheel when it is rotating with a rate of revolution of 30 revolutions per minute.

$$\text{Moment of inertia} = mk^2$$

$$= 300 \times 1.2^2 \quad kg\ m^2$$

$$\text{Angular velocity} = 2\pi f$$

$$= 2\pi \times \frac{30}{60} \quad s^{-1}$$

Hence

$$\text{Kinetic energy} = \tfrac{1}{2}I\omega^2$$

$$= \tfrac{1}{2} \times 300 \times 1.2^2 \times 2\pi \times \frac{30}{60} \quad kg\ m^2 \times s^{-2}$$

$= 680$ J to two significant figures

This is the energy stored in the flywheel. It is the energy that the flywheel can release when it loses its velocity.

FLYWHEELS

Flywheels are devices for storing energy. A flywheel with a moment of inertia I and an angular velocity ω has a kinetic energy of $\tfrac{1}{2}I\omega^2$. Thus the greater the moment of inertia of a flywheel the greater the energy that a rotating wheel will possess for a given angular velocity. For a wheel in the form of a disc the moment of inertia is given by $I = \tfrac{1}{2}mr^2$ Hence the greater the mass and radius of the wheel the greater the moment of inertia and so the greater the energy of the wheel when rotating at some particular angular velocity.

Thus, if a flywheel has a moment of inertia of 100 kg m^2, the energy stored in the wheel when it is revolving at 2 revolutions per second is

$$\text{Kinetic energy} = \tfrac{1}{2}I\omega^2$$

where

$$\omega = 2\pi f$$
$$= 2\pi \times 2\ \text{s}^{-1}$$

Hence

$$\text{Kinetic energy} = \tfrac{1}{2} \times 100 \times (2\pi \times 2)^2 \quad \text{kg m}^2\ \text{s}^{-2}$$
$$= 7900\ \text{J to two significant figures}$$

If the angular speed of the flywheel is doubled, the kinetic energy is quadrupled. Thus, for the flywheel with the moment of inertia of 100 kg m^2, a change in rate of revolution from 2 to 4 revolutions per second results in a change in the kinetic energy of the wheel changing from 7900 J to 31 600 J, i.e. a difference of 23 700 J. A lot of energy is thus needed to change the speed of revolution of the flywheel.

A flywheel can be used to equalize the energy output of an engine and reduce fluctuations of speed that occur. Thus if there is an excess of energy output the energy is 'mopped up' by the flywheel with only a small increase in speed. When there is a drop in output the flywheel can supply a 'top up' of energy without much decrease in speed.

HOISTS

Figure 6.4 shows a simple hoist system, a cable wrapped around a drum. The drum can be rotated by a horizontal shaft and so loads on the free end of the cable lifted.

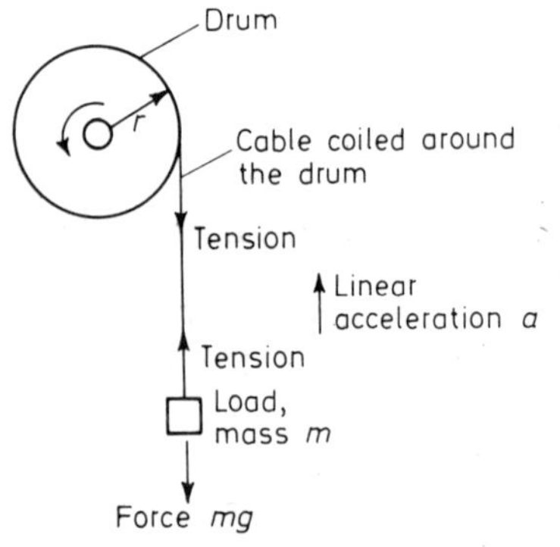

Figure 6.4 A simple hoist

When the load is stationary, and no lifting is occurring, the tension in the cable must be equal to the gravitational force acting on the load, i.e. mg. When however the hoist is operating and the load has an upward acceleration, the resultant force acting on the load is the tension minus the gravitational force, this being the net force which causes the load to accelerate.

When stationary there is no net force on the load, i.e.

$$\text{Tension} = \text{Gravitational force } mg$$

When moving upwards there is a net force on the load,

$$\text{Tension} - \text{gravitational force.}$$

This net force causes a mass m to accelerate with an acceleration a. Hence

$$\text{Tension} - \text{gravitational force} = ma$$
$$\text{Tension} - mg = ma$$
$$\text{Tension} = m(a + g)$$

This cable tension results in a torque being applied to the drum which opposes the applied torque.

$$\text{Net torque on drum} = \text{applied torque} - \text{torque due to cable}$$

But the torque to the tension in the cable is given by

$$\text{Torque due to cable} = \text{tension in cable} \times \text{drum radius}$$
$$= m(a+g) \times r$$

If the applied torque is T then the net torque is

$$\text{Net torque on drum} = T - m(a+g) \times r$$

It is this torque which causes the angular acceleration of the drum. If α is the angular acceleration of the drum and I its moment of inertia about its axis, then

$$\text{Net torque on drum} = I\alpha$$

and so

$$T - m(a+g) \times r = I\alpha$$

or

$$T = I\alpha + mr(a+g)$$

This equation enables us to calculate the torque needed to give a particular linear acceleration to the load.

If the drum starts from rest and accelerates with a constant angular acceleration of α then the angular velocity ω after some time t will be given by

$$\omega = \alpha t$$

and thus the kinetic energy of the drum after this time t will be

$$\text{Kinetic energy of drum} = \tfrac{1}{2}I\omega^2 = \tfrac{1}{2}I\alpha^2 t^2$$

The power required to rotate the drum after a time t will be

$$\text{Power} = T\omega$$

where ω is the angular velocity at that time. As $\omega = \alpha t$ then the power will be

$$\text{Power} = T\alpha t$$

Example 16 A hoist drum has a moment of inertia of 60 kg m^2 and is used to lift a load with a mass of 400 kg. If the upward acceleration of the load is 2 m/s^2 and the drum has a radius of 1 m what will be (a) the torque needed to be applied at the drum, (b) the kinetic energy of the drum after 3 s?

(a)

$$T = I\alpha + mr(a+g)$$
$$a = \alpha r$$

Thus

$$T = \frac{Ia}{r} + mr(a+g)$$
$$= \frac{60 \times 2}{1} + 400 \times 1(2+9.8)$$
$$= 4800 \text{ N m to two significant figures}$$

(b) Kinetic energy $= \frac{1}{2}I\alpha^2 t^2$

$a = \alpha r$

Thus

$$\text{Kinetic energy} = \frac{\frac{1}{2}Ia^2 t^2}{r^2}$$

$$= \frac{60 \times 2^2 \times 3^2}{2 \times 1^2} \qquad \frac{\text{kg m}^2 \times (\text{m/s}^2)^2 \times \text{s}^2}{\text{m}^2}$$

$$= 1080 \text{ J}$$

MOTION IN A CIRCLE

Consider a small object of mass m rotating with a constant speed in a circle path of radius r (Figure 6.5). At point A the velocity will be v in the direction shown. At point B the velocity will still have the same size but be in a different direction. Because the velocity has changed there must have been an acceleration. The velocity has changed because its direction changed. If there had been no change in velocity then the object would have continued in motion in a straight line and would not have deviated from its straight line path to move in a circle.

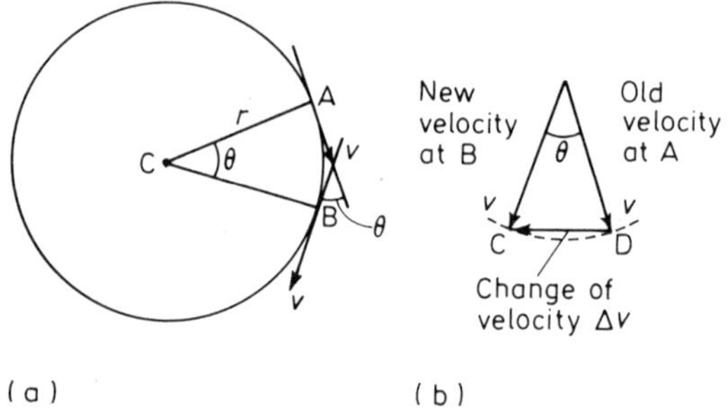

Figure 6.5

The direction of the velocity has to change by an angle θ, this also being the angle subtended at the centre of the circle by the arc AB. The amount by which the velocity has to change is given by the vector triangle in Figure 6.3(b). The change is given by Δv, the amount by which one velocity differs from the other velocity (See Chapter 5 for a discussion of velocity as a vector quantity).

The arc AB, in Figure 6.3(a), is related to the angle θ by

$$\text{arc AB} = r\theta \tag{6.4}$$

If the triangle in Figure 6.3(b) had been an arc of a circle, as indicated by the dotted line, then

$$\text{arc CD} = v\theta$$

If the angle considered had been very small then the arc CD would have been equal to Δv, the change in velocity. If we require the instantaneous acceleration at some point on the circumference of the circle then we need to consider very small angles. Thus we can take arc CD to be equal to Δv in the limit that we require. Hence

$$\Delta v = v\theta \tag{6.5}$$

Eliminating θ between Equations (6.4) and (6.5) gives

$$\frac{\Delta v}{v} = \frac{\text{arc AB}}{r} \tag{6.6}$$

If the time taken for the object to move from A to B is t then this is also the time taken for the velocity change Δv. Thus

$$\text{Acceleration } a = \frac{\Delta v}{t}$$

and using (6.6)

$$a = \frac{\text{arc AB}}{t} \times \frac{v}{r}$$

But as arc AB is the distance covered by the object in time t

$$\frac{\text{arc AB}}{t} = v$$

Hence

$$a = \frac{v^2}{r}$$

The direction of this acceleration is the direction of the Δv, this being directed towards the centre of the circle. Hence the acceleration is called the *centripetal acceleration.*

The above equation gives the acceleration that the object of mass m must experience if it is to move in a circular path with a constant speed. The force necessary to produce this acceleration, the *centripetal force*, must therefore be

$$F = ma$$

$$F = \frac{mv^2}{r}$$

The force will be in the same direction as the acceleration, i.e. directed towards the centre of the circle.

Both the centripetal acceleration and centripetal force equations can be expressed in terms of the angular velocity of the object.

$$v = r\omega$$

Thus

$$a = \omega^2 r$$

and

$$F = m\omega^2 r$$

Example 17 A mass of 0.5 kg is whirled round in a horizontal circle of radius 0.8 m on the end of a cord. What is the tension in the cord when the mass completes 4 revolutions per second?

$$\omega = 2\pi f$$
$$= 2\pi \times 4 \quad \text{rad/s}$$

For the centripetal force in terms of the angular velocity we have

$$F = m\omega^2 r$$
$$= 0.5 \times (8\pi)^2 \times 0.8 \quad \text{kg s}^{-2}\text{ m}$$
$$= 250 \text{ N to two significant figures}$$

Example 18 A 3 kg mass is attached to the end of a cord and whirled round in a vertical circle of radius 1 m. What is the maximum tension in the cord when the mass completes 3 revolutions per second?

$$\omega = 2\pi f$$
$$= 2\pi \times 3 = 6\pi \quad \text{rad/s}$$
$$F = m\omega^2 r$$
$$= 3 \times (6\pi)^2 \times 1 \quad \text{kg s}^{-2}\text{ m}$$
$$= 1066 \text{ N to four significant figures}$$

This is the force needed to keep the object moving in a circle. The forces acting on the mass are the tension in the cord and the gravitational force due to the mass, i.e. 3 × 9.81 N. When the mass is at the top of its vertical path the tension in the cord and the gravitational force are both acting in the same direction. Thus at the top of path

$$\text{Tension} + 3 \times 9.81 = 1066 \text{ N}$$
$$\text{Tension} = 1036 \text{ N}$$

When the mass is at the bottom of its path the tension in the cord is in the opposite direction to the gravitational force. Thus at the bottom of path

$$\text{Tension} - 3 \times 9.81 = 1066 \text{ N}$$
$$\text{Tension} = 1095 \text{ N}$$

This is the maximum tension.

VEHICLES ROUNDING CURVES

For a vehicle to turn a corner with a constant speed there must be a radial force giving an acceleration so that the velocity of the vehicle changes. The force for a vehicle of mass m moving with a speed v round a corner of radius r must be mv^2/r. If the road surface was a horizontal sheet of ice then it is unlikely that the vehicle would be able to turn the corner but would instead continue in a straight line. This is because the force to turn the corner comes from the friction force between the tyres of the car and the road surface. If the frictional force is not large enough, i.e. equal to mv^2/r, then skidding occurs as the vehicle tends to continue in its straight line path.

In order not to rely on friction the corner can be banked. Figure 6.6 shows the general arrangement. With banking there is a component of the reactive force that acts on the vehicle which is able to provide the centripetal force necessary for the vehicle to move in a circular path. The component of the reactive force that is horizontal is $R \sin \theta$,

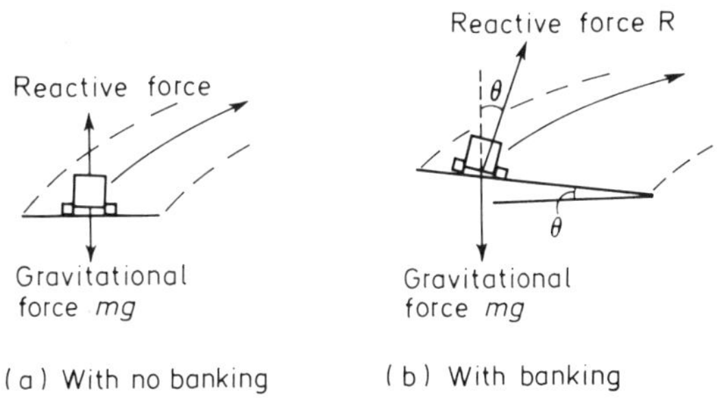

Reactive force

Gravitational force mg

(a) With no banking

Reactive force R

Gravitational force mg

(b) With banking

Figure 6.6

where θ is the angle at which the corner is banked. The vertical component of the reactive force must be equal to the gravitational force mg that acts on the vehicle. Thus

$$R \sin \theta = \frac{mv^2}{r}$$

if the entire centripetal force is to be provided by the reactive force component. For the vertical reactive component

$$R \cos \theta = mg$$

Dividing the two equations gives

$$\tan \theta = \frac{v^2}{rg}$$

Thus for the value of v given by a particular banking angle and curve radius there is no need of friction to provide any part of the centripetal force. A curve can only be banked for a particular speed. Designers generally arrange the banking of a road for the average speed of the traffic at that corner. The equation also gives the angle at which an aircraft has to bank in order to execute a turn.

Example 19 Calculate the maximum speed with which a vehicle can traverse an unbanked curve of radius 40 m without skidding. The limiting coefficient of sliding friction between the tyres and road can be taken as 0.6.

Gravitational force acting on the vehicle $= mg$

Hence

Reactive force $= mg$

Thus

Frictional force $= \mu mg$

where μ is the coefficient of friction (see Chapter 5). Hence when the centripetal force exceeds this frictional force, skidding will occur. The maximum speed must therefore be when the centripetal force equals the frictional force.

$$\frac{mv^2}{r} = \mu mg$$

$$v^2 = \mu rg$$

$$v^2 = 0.6 \times 40 \times 9.8 \qquad \text{m} \times \text{m/s}^2$$

v = 15 m/s to two significant figures.

Example 20 Calculate the angle of banking required on a corner of radius 60 m so that vehicles can travel round the corner at 60 m/s without any side thrust on the tyres.

Without any side thrust means that the vehicle must traverse the corner without relying on frictional forces. Thus

$$\tan\theta = \frac{v^2}{rg}$$

$$= \frac{60^2}{60 \times 9.8} \qquad \frac{(\text{m/s})^2}{\text{m} \times \text{m/s}^2}$$

$$= 6.12$$

$$\theta = 80.7^\circ$$

PROBLEMS

1. A flywheel rotates through 5 complete revolutions in 2 s. What is the average angular velocity over that time interval?

2. The angular velocity of a flywheel changes from 4 s^{-1} to 10 s^{-1} in 3 s. What is the average angular acceleration over that time interval?

3. A flywheel is rotating at 12 revolutions per minute. What is its angular velocity?

4. A grinding wheel rotates at 3000 revolutions per minute. What is its angular velocity?

5. A spindle starts from rest and attains a frequency of rotation of 600 revolutions per minute in 20 s. Calculate the angular acceleration if it is assumed to be uniform over the time interval concerned.

6. A flywheel slows down from 500 revolutions per minute to 300 revolutions per minute in 2 minutes. What was the angular acceleration if it can be assumed to have been constant over the time concerned?

7. A flywheel increases its frequency of rotation from 20 to 40 revolutions per minute in 10 s with a uniform acceleration. Calculate (a) the angular acceleration and (b) the number of revolutions made during the 10 s period.

8. A rotating body is rotating with an angular velocity of 5 s^{-1} when it is accelerated at 2 s^{-2}. What will be the angular velocity after 5 s of this acceleration?

9. A grinding wheel starts from rest and accelerates uniformly with an angular acceleration of 3 s^{-2}. What will be its angular velocity after 6 s and the total angle rotated in that time?

10. A car engine is idling at 600 revolutions per minute. When the fuel flow to the engine is increased by the accelerator pedal being depressed the number of revolutions per minute increases to 3000 per minute in

5 s. If the angular acceleration was constant over the time interval concerned what was (a) the initial angular velocity, (b) the angular velocity after the 5 s, (c) the angular acceleration, (d) the number of revolutions made during the 5 s?

11. A grinding wheel has a diameter of 200 mm. Calculate the linear speed of a point on the rim of the wheel when it is rotating at 3000 revolutions per minute.

12. A flywheel accelerates from 10 to 30 revolutions per minute in 10 s. Calculate the angular acceleration and the linear acceleration for a point on the rim if the wheel has a diameter of 2 m.

13. The linear speed of a belt that passes round a pulley wheel is 30 m/s. If there is no slip what will be the angular velocity of the 200 mm diameter pulley?

14. What torque is required to cause a pulley to rotate at 8 revolutions per second after 10 s starting from the pulley being at rest? The pulley has a mass of 20 kg and a radius of gyration of 650 mm.

15. A flywheel has a mass of 400 kg and a radius of gyration of 0.7 m. If the flywheel is rotating at 500 revolutions per minute calculate the constant torque needed to bring it to rest in 30 s and how many revolutions the wheel will make before coming to rest.

16. Calculate the moment of inertia of a solid disc 300 mm in diameter and 25 mm thick about its axis. The disc is made of material with a density of 7700 kg/m^3.

17. The rotor of an electric motor has a mass of 100 kg and a radius of gyration of 80 mm. Calculate the torque required to accelerate the rotor at 0.3 s^{-2}. At such an acceleration how long will it take to reach 10 revolutions per second starting from rest?

18. Calculate the moment of inertia of a rod 40 mm in diameter and 2 m long, mass 7 kg, about an axis perpendicular to the rod and passing through its centre.

19. Calculate the torque required to raise the speed of revolution of a winding drum from 30 to 60 revolutions per minute in 60 seconds if there is a frictional torque of 14 kN m. The winding drum has a mass of 200 tonne and a radius of gyration of 3 m.

20. Calculate the uniform torque needed to accelerate the rotating table of a vertical boring machine to 1 revolution per second in four complete revolutions starting from rest. The table has a mass of 600 kg and a radius of gyration of 650 mm.

21. Calculate the power developed when a constant torque of 100 Nm gives a rotation of 360 revolutions per minute.

22. The shaft of an electric motor rotates at 1200 revolutions per minute when the output is 3 kW. Calculate the output torque of the motor.

23. A constant torque of 120 N m is used to keep a flywheel rotating at a constant 6 revolutions per second. What is the power input to the wheel?

24. Design a flywheel capable of storing 100 000 J of energy when rotating at 500 revolutions per minute.

25. A flywheel has a mass of 250 kg and a radius of gyration of 1 m. Calculate the energy stored in the flywheel when it is rotating at 50 revolutions per minute. How would the energy stored change if the rate at which the flywheel revolved was doubled?

26. Calculate the gain in kinetic energy by a flywheel when it is accelerated from 200 to 260 revolutions per minute. The flywheel has a mass of 70 kg and a radius of gyration of 500 mm.

27. The drum of a hoist of radius 1 m has a mass of 120 kg and a radius of gyration of 400 mm. If the drum is used to raise a mass of 800 kg with an acceleration upwards of 1.6 m/s^2, calculate the torque required at the drum and the power required after accelerating the load for 2 s from rest.

28. The drum of a hoist has a moment of inertia of 80 kg m^2 and a diameter of 1 m. Determine the torque required at the drum to raise a load of mass 1000 kg with an upward acceleration of 2 m/s^2. Calculate the power required after accelerating for 4 s from rest.

29. A load of mass 3000 kg is raised with a uniform acceleration of 1.5 m/s^2 by means of a light cable passing over the drum of a hoist. The drum has a diameter of 1.5 m, a mass of 2000 kg and a radius of gyration of 800 mm. What is the torque required and the power required after 3 seconds from rest?

30. The flywheel of an engine is required to provide 500 J of energy when its speed changes from 500 revolutions per minute to 400 revolutions per minute. What moment of inertia is required for the flywheel?

31. Explain the function of a flywheel.

32. A mass of 0.3 kg is whirled round in a horizontal circle of radius 0.6 m on the end of a cord. What is the tension in the cord when the mass completes 5 revolutions per second? How will the tension change if the angular speed is doubled?

33. A mass of 0.6 kg is whirled round in a vertical circle of radius 0.8 m at the end of a cord. What is the maximum and the minimum tension in the cord when the mass completes 200 revolutions per minute?

34. Explain what would happen to the mass in Problem 32 if the string suddenly broke.

35. Explain what would happen to the mass in Problem 33 if the string suddenly broke.

36. Calculate the greatest speed with which a car may pass over a hump back bridge without leaving the ground if the bridge forms the arc of a vertical circle of radius 8 m.

37. Calculate the angle of banking required on a corner of radius 80 m so that vehicles can travel round the corner at 40 m/s without any side thrust on the tyres.

38. Calculate the maximum speed with which a vehicle can travel round a corner, unbanked and horizontal, of radius 50 m without skidding. The limiting coefficient of sliding friction between the tyres and the road can be taken as 0.6. If rain reduces the coefficient to 0.2, how is the speed changed?

7 Simple harmonic motion

OSCILLATIONS

Figure 7.1 shows two forms of oscillator. In Figure 7.1(a) a mass on the end of a tethered spring is pulled down and then released. It bobs up and down, first to one side of the initial rest position of the mass then the other. The oscillation, the bobbing up and down, can continue for some time. In Figure 7.1(b) there is a simple pendulum. This is a mass at the end of a string, the upper end of the string being fixed to some rigid support. When the mass is pulled to one side and then released it swings from side to side, first to one side of the initial rest position of the mass then the other. The oscillation, the swinging back and forth, can continue for some time. Both the oscillations have a displacement/time graph that looks something like Figure 7.1(c).

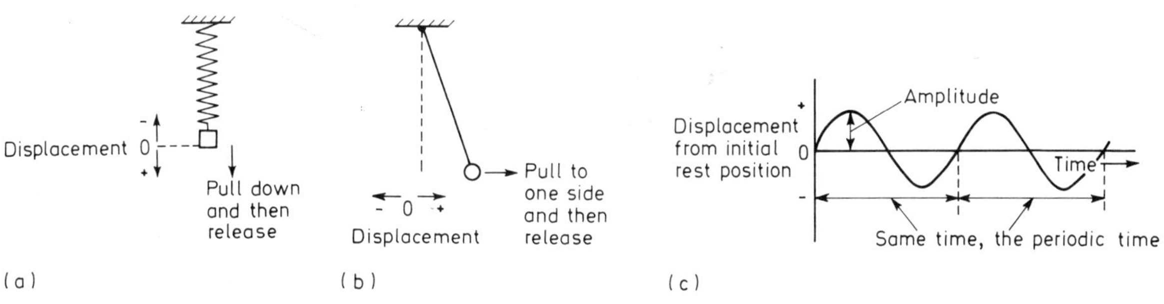

Figure 7.1 Oscillations

The motion is confined to within a range $\pm A$, where A represents the displacement called the amplitude in Figure 7.1(c). Each to and fro movement takes place in the same time. This is true despite the amplitude of the oscillations decreasing with time due to a loss of energy from the system due to frictional effects. With both the mass on the spring and the mass on the cord there is a force acting which endeavours to restore the mass to its initial rest position.

Any sort of motion which repeats itself in equal intervals of time is called *periodic*. The *periodic time* is the time taken for one complete to and fro motion. For a periodic oscillation this time is a constant regardless of the amplitude of the oscillation. The *frequency* of the oscillations is the number of complete vibrations per second, this being the reciprocal of the periodic time.

$$\text{Frequency}, f = \frac{1}{T}$$

UNIT: Hertz (Hz)

A frequency of 50 Hz means 50 complete oscillations every second. Thus the time for one complete oscillation is 1/50 s. Instead of the term the hertz, the term cycles per second used to be used.

SIMPLE HARMONIC MOTION

The oscillating spring and the simple pendulum in Figure 7.1 are both shown as giving the same form of displacement/time graph. A large number of oscillating objects give a similar graph.

Figure 7.2(a) shows part of the displacement/time graph for such an oscillation. The graph represents the motion from one extreme displacement to the other extreme. As velocity is the rate of change of displacement with time then the velocity at any instant must be the slope of the displacement/time graph at that instant. By determining the slopes at a number of points on the displacement/time graph a velocity/time graph can be obtained (Figure 7.2(b)). The maximum velocity occurs when the slope of the displacement/time graph is a maximum and this is when the oscillating object is passing through its rest or zero displacement position. The velocity is zero where the slope of the displacement/time graph is zero and this is at the maximum displacement positions. Acceleration is the rate of change of velocity with time and so the acceleration at any instant must be the slope of the velocity/time graph at that instant. By determining the slopes at a number of points on the velocity/time graph an acceleration/displacement graph can be produced. The graph is a straight line (Figure 7.2(c)). The acceleration at any instant is directly proportional to the displacement at that instant.

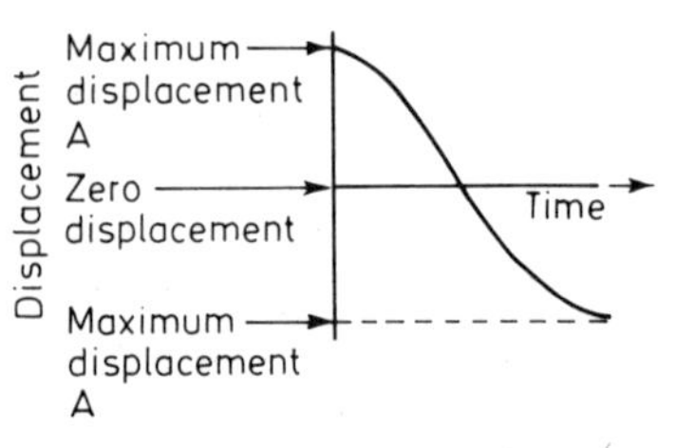

(a) Displacement / time graph

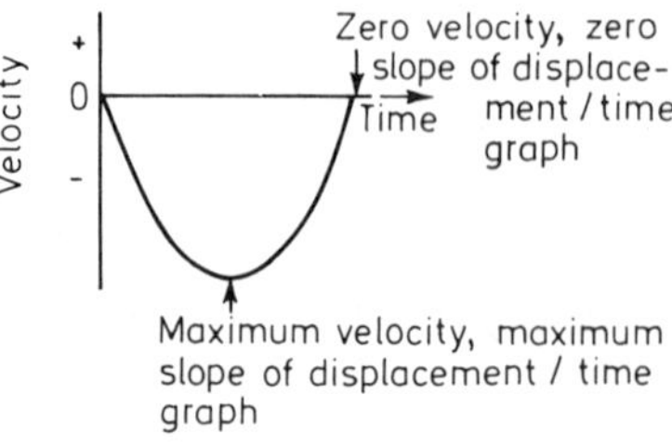

(b) Velocity / time graph

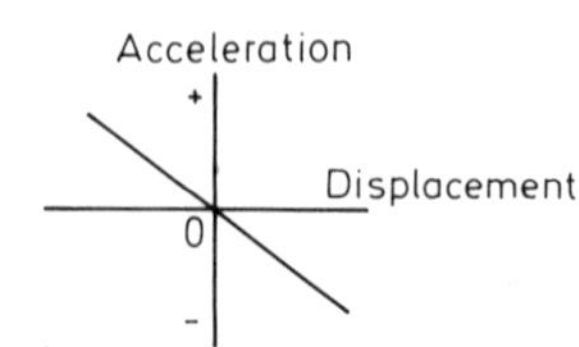

(c) Acceleration/displacement graph

Figure 7.2

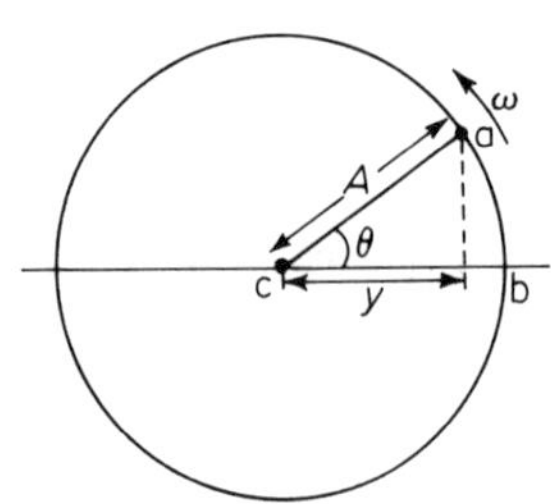

Figure 7.3

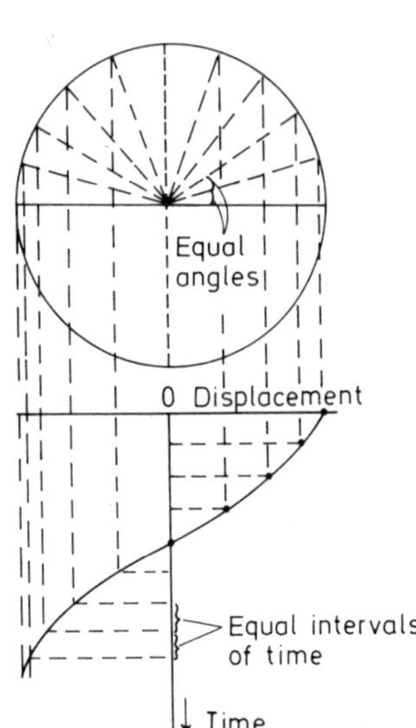

Figure 7.4

The form of the displacement/time graph is that of a cosine graph. This means that the displacement y can be calculated from the following equation

$$y = A \cos \theta$$

where A is the amplitude and θ is proportional to the time. This equation also describes another type of motion, the projection of a rotating arm (called a *phasor*) moving with constant angular velocity. Figure 7.3 illustrates this with Figure 7.4 showing how the projection varies with time for a uniform angular velocity. If the angular velocity is ω then the angle θ moved in a time t is given by

$$\theta = \omega t$$

We can therefore consider an oscillator to have a displacement/time relationship described by the equation

$$y = A \cos \omega t$$

The velocity at any instant of time is the rate of change of displacement with time. This is $\mathrm{d}y/\mathrm{d}t$. Differentiating the above displacement equation gives

$$\frac{\mathrm{d}y}{\mathrm{d}t} = -A\omega \sin \omega t$$

The velocity/time graph is thus described by the equation

$$v = -A\omega \sin \omega t$$

This agrees with the graph given in Figure 7.2(b), which is a sine graph.

The acceleration at any instant is the rate of change of velocity with time. This is $\mathrm{d}v/\mathrm{d}t$. Differentiating the velocity equation gives

$$\frac{\mathrm{d}v}{\mathrm{d}t} = -A\omega^2 \cos \omega t$$

$$a = -A\omega^2 \cos \omega t$$

But $y = A \cos \omega t$, hence

$$a = -\omega^2 y$$

The acceleration is proportional to the displacement, as in Figure 7.2(c).

This acceleration equation tells us that when $y = 0$ the acceleration is zero. Thus there is no acceleration when the pendulum bob is passing through the rest position. When y is at its maximum value then the acceleration has its biggest value. This value is however negative. A negative acceleration means that the velocity is decreasing. If the sign had been positive then the bob would have been speeding up, its velocity increasing. The bob has zero velocity at its maximum displacement.

The velocity equation involves the sine of the angle.

$$v = -A\omega \sin \omega t$$

$$v = -A\omega \sin \theta$$

where θ is the angle covered in time t. This is the angle indicated in Figure 7.3. In that figure

$$\sin \theta = \frac{ab}{ac}$$

But, using Pythagoras' theorem,

$$ac^2 = ab^2 + bc^2$$

or

$$A^2 = ab^2 + y^2$$

$$ab^2 = A^2 - y^2$$

$$ab = \sqrt{(A^2 - y^2)}$$

Hence

$$\sin \theta = \frac{\sqrt{(A^2 - y^2)}}{A}$$

Hence

$$v = -A\omega \times \frac{\sqrt{(A^2 - y^2)}}{A}$$

$$v = -\omega\sqrt{(A^2 - y^2)}$$

The maximum velocity thus occurs when y is zero.

$$\text{Maximum velocity} = -\omega A$$

An acceleration indicates that a resultant force is acting on an object. Thus as $F = ma$ the force acting on the pendulum bob during its oscillation must be

$$F = -m\omega^2 y$$

The force is thus proportional to the displacement.

The motion can thus be described by the following:

1. The displacement/time relationship is a cosine function.

$$y = A \cos \omega t$$

2. The velocity/time relationship is a sine function.

$$v = -A\omega \sin \omega t$$

3. The acceleration is proportional to the displacement.

$$a = -\omega^2 y$$

4. The restoring force is proportional to the displacement.

$$F = -m\omega^2 y$$

5. The acceleration and the restoring force are always directed towards a fixed point in the path, the zero displacement point.

Though the above might be obtained from a consideration of the motion of a pendulum, one for which there was no noticeable frictional effects, the results are general enough to be applicable to a wide range of oscillations. Oscillations for which the above are valid are said to perform *simple harmonic motion.*

Instead of describing the oscillations in terms of an angular velocity the frequency of the oscillation is often used. If T is the time for one complete oscillation, i.e. the periodic time, then for the motion of the particle round the circle (Figure 7.3) there will have been one complete rotation. This is an angular rotation of 2π radians in time T. Thus

$$\omega = \frac{2\pi}{T}$$

But

$$f = \frac{1}{T}$$

and so

$$\omega = 2\pi f$$

Example 1 A body moving with simple harmonic motion has an acceleration of 1.5 m/s^2 when 200 mm from the zero displacement position. Calculate the frequency of the oscillation.

$$a = -\omega^2 y$$

$$a = -(2\pi f)^2 y$$

Therefore

$$f = \frac{1}{2\pi}\sqrt{\frac{a}{y}}$$

$$= \frac{1}{2\pi}\sqrt{\frac{1.5}{0.200}}$$

$$= 2.2 \text{ Hz to two significant figures}$$

Example 2 Calculate the time taken for one complete oscillation of the oscillator in Example 1.

$$T = \frac{1}{f}$$

$$= \frac{1}{2.2}$$

$$= 0.46 \text{ s to two significant figures}$$

Example 3 A piston moves with simple harmonic motion. If the amplitude of the piston oscillation is 70 mm and its frequency is 10 Hz calculate the maximum acceleration experienced by the piston.

$$a = -\omega^2 y$$

The maximum value will occur when y is a maximum, i.e. equal to the amplitude.

$$\omega = 2\pi f$$

Hence

$$a = -(2\pi f)^2 y$$

$$= -(2\pi \times 10)^2 \times 0.070 \qquad \text{s}^{-2} \times \text{m}$$

$$= -280 \text{ m/s}^2 \text{ to two significant figures}$$

THE SIMPLE PENDULUM

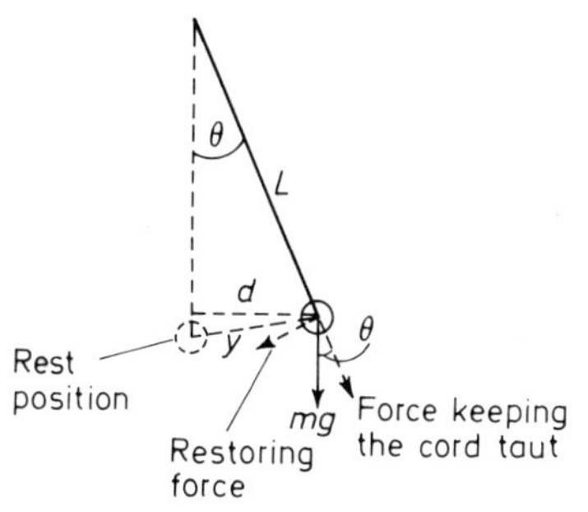

Figure 7.5 The simple pendulum

With a simple pendulum the force endeavouring to restore the pendulum bob back to its initial rest position is gravity (Figure 7.5). The restoring force is the component of the gravitational force that acts at right angles to the pendulum cord. The gravitational force acting on a pendulum bob of mass m is mg and the component acting as the restoring force is $mg \sin \theta$. The other component $mg \cos \theta$ is the force keeping the cord taut and so equates to the tension in the cord.

$$\text{Restoring force} = -mg \cos \theta$$

$$\cos \theta = \frac{d}{L}$$

If the arc through which the pendulum swings is small then d and y are virtually the same. With this approximation

$$\text{Restoring force} = \frac{-mgy}{L}$$

But the restoring force for simple harmonic motion should be proportional to the displacement. This is correct as the above equation indicates, when we make the approximation. The motion of a simple pendulum is thus only reasonably simple harmonic for small angles of oscillation.

The restoring force for simple harmonic motion is given by

$$F = -m\omega^2 y$$

Hence, for the simple pendulum

$$-m\omega^2 y = \frac{-mgy}{L}$$

$$\omega^2 = \frac{g}{L}$$

As $\omega = 2\pi f$ then

$$f = \frac{1}{2\pi}\sqrt{\frac{g}{L}}$$

The frequency of oscillation of a simple pendulum thus depends on the value of the acceleration due to gravity and the length of the pendulum. The frequency does not depend on the amplitude of the oscillation.

Example 4 Calculate the length of cord needed for a simple pendulum to have a periodic time of 2 s.

$$\text{Periodic time} = \frac{1}{f}$$

Hence

$$f = ½\text{ Hz}$$

$$f = \frac{1}{2\pi}\sqrt{\frac{g}{L}}$$

$$\frac{1}{2} = \frac{1}{2\pi}\sqrt{\frac{9.8}{L}}$$

Hence

$$L^2 = \frac{\pi}{\sqrt{9.8}}$$

$$L = 3.2\text{ m to two significant figures}$$

MASS ON A VERTICAL SPRING

A mass suspended from a vertical spring can be made to oscillate by pulling the mass down, so extending the spring, and then releasing. The restoring force acting on the mass is the force resulting from the extension of the spring. The force is proportional to the extension of the spring if it obeys Hooke's law.

$$F \propto y$$

or

$$F = ky$$

where k is a constant, the force required to produce unit extension of the spring, sometimes referred to as the *stiffness* of the spring.

$$\text{Restoring force} = -ky$$

The restoring force is thus proportional to the displacement for a spring for which Hooke's law holds. This is the condition for simple harmonic motion and so the oscillating mass must oscillate with simple harmonic motion.

The restoring force for simple harmonic motion is given by

$$F = -m\omega^2 y$$

Hence

$$-m\omega^2 y = -ky$$

$$\omega^2 = \frac{k}{m}$$

As $\omega = 2\pi f$, then

$$f = \frac{1}{2\pi}\sqrt{\frac{k}{m}}$$

The frequency of oscillation of the mass thus depends on the stiffness of the spring and the mass of the object on the end of the spring. It does not depend on the amplitude.

Example 5 A load is suspended from a vertically mounted spring and causes it to extend by 10 mm. With what frequency will the spring oscillate if the load is pulled further down and then released?

The force causing the initial extension of 10 mm is mg, where m is the mass of the load.

$$F = ky$$

Thus

$$mg = ky$$

$$k = \frac{mg}{y}$$

$$k = \frac{9.8\text{ m}}{0.010}$$

Hence

$$f = \frac{1}{2\pi}\sqrt{\frac{k}{m}}$$

$$= \frac{1}{2\pi}\sqrt{\frac{9.8\text{ m}}{0.010\text{ m}}}$$

$= 5.0$ Hz to two significant figures

Example 6 A spring of stiffness 15 kN/m supports a load of 20 kg. With what frequency will the load oscillate and what will be the maximum acceleration if the amplitude of the oscillation is 15 m?

$$f = \frac{1}{2\pi}\sqrt{\frac{k}{m}}$$

$$= \frac{1}{2\pi}\sqrt{\frac{15 \times 10^3}{20}}$$

$= 4.4$ Hz to two significant figures

The maximum acceleration is at the maximum displacement. Thus

$$\text{Maximum acceleration} = -\omega^2 y = -(2\pi f)^2 y = -(2\pi \times 4.4)^2 \times 0.015 = -7.6\ \text{m/s}^2 \text{ to two significant figures}$$

RESONANCE

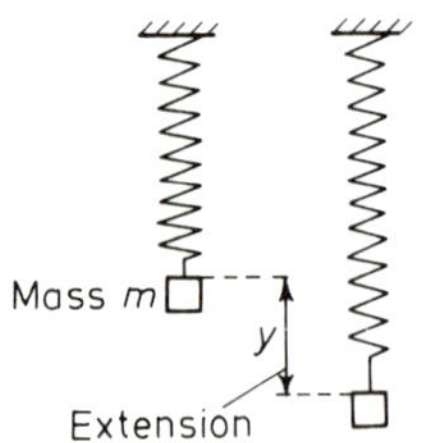

Figure 7.6 A mass on a vertical spring

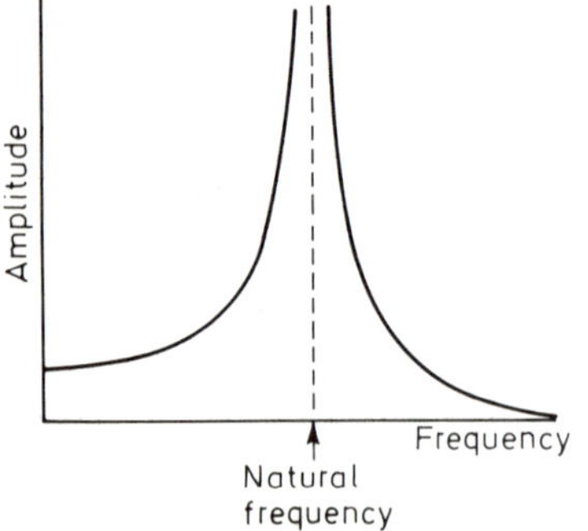

Figure 7.7

Suppose we have a vertical spring mounted as in Figure 7.6, a mass being attached to its lower end. So far the oscillations of such a spring have been considered in terms of those induced by pulling the mass down and then releasing it. But what happens to such a system if the support to which the upper end of the spring is attached vibrates? Figure 7.7 shows how the oscillation amplitude produced for the mass on the spring depends on the frequency of the vibration of the support. At one particular frequency the amplitude of the mass is considerably greater than at any other frequency. This frequency is in fact the frequency with which the mass would oscillate if pulled down and allowed to oscillate without any vibration from the support, this frequency being called the *natural frequency* of the system. When the applied frequency equals the natural frequency the large amplitude oscillation is said to be produced because of resonance. *Resonance* occurs when the applied frequency to a system equals the natural frequency of that system.

If you ever have had to push a child on a swing you will have quickly realised that the way to build up a large amplitude oscillation is to push the swing in a regular, repetitive way. The frequency of your pushes has to equal the frequency with which the swing is oscillating for the large amplitudes to be produced. Your pushes have to be correctly timed. If you try pushing the swing with a frequency different to the 'natural frequency' of the swing then no large amplitude is produced.

Resonance is a problem in industry. There are many sources of vibration and as all systems have at least one resonant frequency there is always the possibility of the vibration being at or close to the resonant frequency and so resulting in a relatively large amplitude oscillation. Such an oscillation may be completely undesirable. A simple example of this is in machine tools. If the tool is being used for accurate work a vibration which results in large amplitude oscillations of the tool will certainly be undesirable.

PROBLEMS

1. For simple harmonic motion describe the relationships between the following: (a) the displacement and the time, (b) the velocity and the time, (c) the acceleration and the displacement, (d) the restoring force and the displacement.

2. Describe the characteristics of simple harmonic motion.

3. An oscillator takes 4 s to complete one oscillation. What is the frequency of the oscillation?

4. A body oscillates with simple harmonic motion. The frequency of the oscillations is 12 Hz and the amplitude 200 mm. What is the maximum acceleration and the maximum velocity attained and at what points in the path of the oscillator are these maxima attained?

5. A valve moves with simple harmonic motion. If the amplitude of the movement is 5 mm and the periodic time 0.005 s what will be (a) the maximum velocity, (b) the maximum acceleration?

6. A body with a mass of 2 kg oscillates in a straight line with simple harmonic motion. The maximum restoring force applied to the mass is 200 N and the amplitude of the oscillation 800 mm. What is (a) the frequency, (b) the time for one oscillation, (c) the maximum acceleration?

7. A body with a mass of 20 kg oscillates in a straight line with simple harmonic motion. The motion has a frequency of 2 Hz and the maximum velocity attained by the body is 4 m/s. What is (a) the amplitude of the motion, (b) the maximum acceleration?

8. Part of a machine oscillates with simple harmonic motion. The frequency of the oscillation is 3 Hz and its amplitude is 200 mm. What is the maximum acceleration? If the part concerned has a mass of 4 kg what is the maximum restoring force?

9. Calculate the length of a simple pendulum required for a periodic time of 0.5 s.

10. A vertical helical spring has a stiffness of 2 kN/m. Calculate the frequency of the oscillation that will be produced when a mass of 3 kg is supported by the spring.

11. A vertical helical spring has a stiffness of 900 N/m. Calculate the frequency of the oscillation that will be produced when a mass of 2 kg is supported by the spring. What will be the maximum acceleration and at what point in the oscillation will it occur when the amplitude is 30 mm?

12. A 2.5 kg mass when attached to the lower end of a vertical helical spring causes the spring to extend by 20 mm. What will be the frequency of oscillation if the mass is made to oscillate at the end of the spring?

13. When a machine is mounted on an assembly of vertical springs each spring is compressed by 20 mm. With what frequency will the mass on the spring oscillate?

14. A vertical helical spring extends by 15 mm when a load of 200 g is attached to the lower end of the spring. With what frequency will the spring oscillate?

15. Explain what is meant by resonance.

16. A vertical helical spring with a stiffness of 500 N/m has a mass of 800 g attached to its lower end. The upper end of the spring is attached to a support. If the support is vibrating with what frequency will the spring–mass system resonate?

17. If the mass attached to the spring in Problem 16 was doubled how would the resonant frequency change?

18. A machine is mounted on springs. How will the resonant frequency change if (a) the springs are made stiffer (b) more mass is attached to the machine?

8 Fluids in motion

FLUID PRESSURE

A *fluid* can be defined as anything that is capable of flowing. Thus a liquid and a gas are fluids. Fluids exert pressures. If an area A of a container surface experiences a force F as a result of the fluid in the container then we can state that there is a fluid *pressure* acting on that surface of F/A.

$$\text{Pressure} = \frac{\text{force}}{\text{area}}$$

UNIT: pascal (Pa); 1 Pa = 1 N/m^2.

Thus if a container surface of area 2 m^2 experiences a force of 100 N then the pressure of the fluid on that surface is

$$\text{pressure} = \frac{100}{2} \quad \frac{\text{N}}{\text{m}^2}$$

$$= 50 \text{ N/m}^2$$

$$= 50 \text{ Pa.}$$

The force exerted by a fluid on a surface is often called the *thrust*. Thus in the above example the thrust on the surface due to the fluid is 100 N.

The pressure in a fluid has certain features:

1. The pressure exerted on any point on a surface in the fluid is always at right angles to the surface.
2. The pressure at any point in a fluid is the same in all directions at that point.

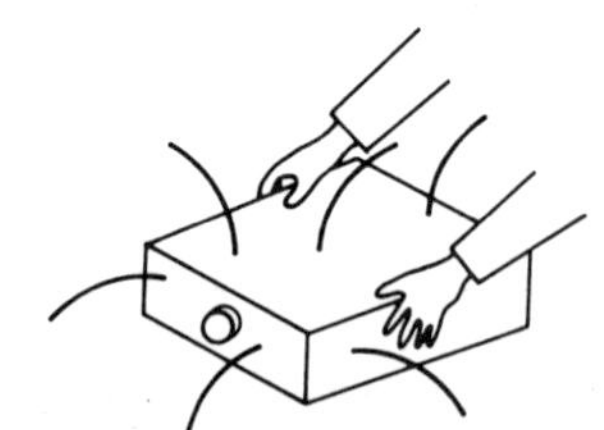

Figure 8.1

If you take a polythene container full of water (Figure 8.1) and press on the container the water will squirt out if the container has holes in its surfaces. The water always emerges at right angles to the surface, regardless from where in the surface the hole is. The water also squirts out of all the holes. It comes out in all directions.

Fluids have a weight and if they are in a container they must exert a pressure on the base of the container. Consider the case shown in Figure 8.2 and the pressure due to a height h of liquid above the base of the container. The mass of liquid above the base is the volume of liquid multiplied by the density, i.e. $hA\rho$. The gravitational force thus acting on the base due to the liquid is therefore $hA\rho g$. This is the force acting over an area of A. Hence the force per unit area, i.e. the pressure is

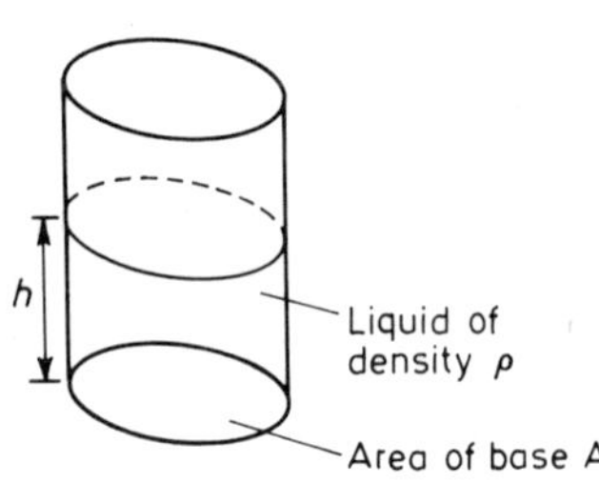

$$\text{Pressure} = \frac{hA\rho g}{A}$$

$$= h\rho g$$

The equation gives the *gauge pressure*. This is the pressure above that due to the atmosphere acting on the surface. A simple *manometer* (Figure 8.3) can be used to measure gauge pressure. The total pressure

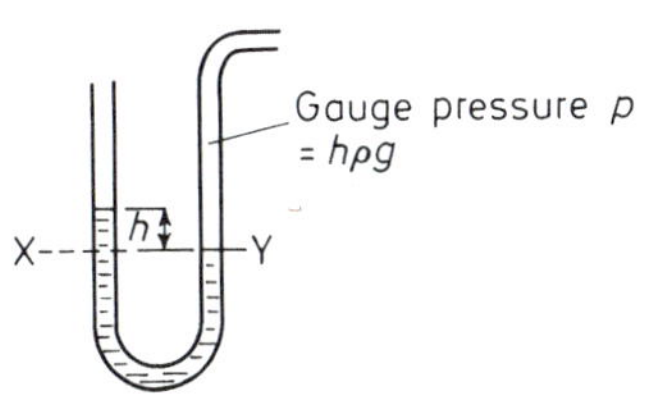

Figure 8.3 Simple manometer

is called the *absolute pressure*. The pressure due to the atmosphere above the earth's surface is about 101 kPa.

Absolute pressure = gauge pressure + atmospheric pressure

Example 1 Calculate the pressure difference in pascals between the two limbs of a U-tube manometer for which the height difference is 20 mm of mercury. The density of mercury is 13.6×10^3 kg/m³.

$$\text{Pressure difference} = h\rho g$$
$$= 0.020 \times 13.6 \times 10^3 \times 9.81$$
$$= 2.67 \text{ kPa}$$

STEADY FLOW

When a fluid flows along a pipe or channel the flow is said to be steady when the velocity and pressure of the fluid at any particular section of the pipe or channel is not varying with time. Figure 8.4 illustrates this. For the section XX the pressure and velocity are unchanging with time. Similarly for the section YY. For Figure 8.4(a) with a uniform tube the conditions at XX are identical with those at YY and thus we would expect the pressure and fluid velocity to be the same for both. For Figure 8.4(b) the tube is tapered and thus though the velocity and pressure at XX and that at YY does not vary with time they would not be expected to be the same.

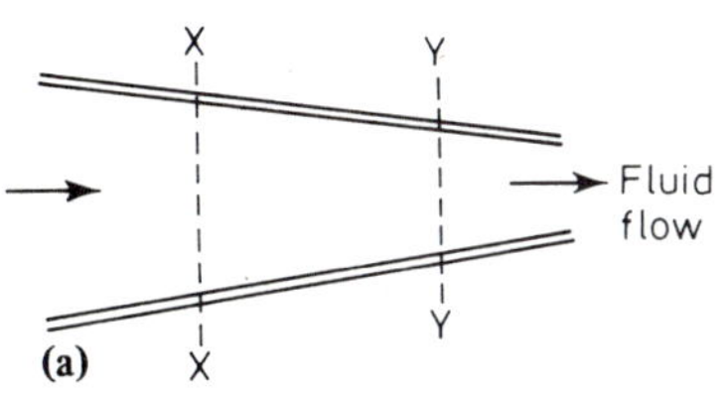

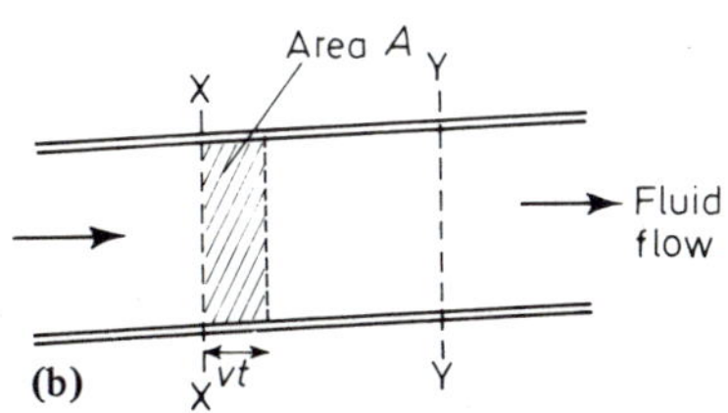

Figure 8.4

If the velocity of the fluid flow in Figure 8.4(a) is constant with a value v then the distance covered by any part of that fluid in a time t must be vt. The volume of fluid that moves past XX in time t must be therefore $A \times vt$. Thus

$$\text{Volumetric rate of flow } \dot{V} = Av$$

where the *volumetric rate of flow* is the volume flowing through a section per second.
UNIT: m³/s.

In the case of the tapered tube the volumetric rate of flow at section XX will be A_1v_1, where A_1 is the cross-sectional area at XX and v_1 the velocity of the fluid at XX. The volumetric rate of flow at section YY will be A_2v_2 where A_2 is the cross-sectional area at YY and v_2 the velocity of the fluid at YY. The quantity of fluid passing XX per second must equal the quantity passing YY per second, thus

$$A_1v_1 = A_2v_2$$

This is known as the *equation of continuity*.

The above equation applies if the fluid is incompressible. If the fluid is compressible the volume passing XX per second may be different from the volume passing YY per second. The mass passing XX per second will however equal the mass passing YY per second. Thus

$$\rho_1A_1v_1 = \rho_2A_2v_2$$

where ρ_1 is the fluid density at XX and ρ_2 the density at YY. The mass flowing past a section per second is called the *mass flow rate*.

$$\text{Mass flow rate } \dot{M} = \rho\dot{V}$$

UNIT: kg/s.

The above discussions have assumed that the velocity of the fluid flow along a pipe is the same across an entire section, i.e. from one wall of the pipe to the other. This is not correct. The velocity does vary across the pipe. If the flow is orderly then the velocity across most of the pipe section is constant, only dropping close the tube walls. Thus in practice the velocities v_1 and v_2 in the equation of continuity represent the mean velocity across the section. The mean velocity can be determined from a measurement of the volumetric rate of flow through a pipe of known cross-sectional area and use of the equation $\dot{V} = Av$.

Example 2 Calculate the quantity of water delivered per second through a nozzle of cross-sectional area 1000 mm² if the water has a velocity of 20 m/s when passing through the nozzle.

$$\dot{V} = Av$$

$$= 1000 \times 10^{-6} \times 20 \qquad \mathrm{m^2 \times m/s}$$

$$= 20 \times 10^{-3}\ \mathrm{m^3/s}$$

ENERGY OF A FLUID IN MOTION

If you push a mass against a spring and compress it, or against any force, then work is done. Similarly if you push fluid against the pressure of the fluid then work is done. When a fluid flows through a pipe against the pressure in the pipe then work is done. Thus, if at some section of a pipe the pressure is p, then the force opposing the motion of the fluid is given by

$$\text{Pressure} = \frac{\text{force}}{\text{area}}$$

where A is the cross-sectional area at the section concerned.

The work done in moving a distance L against the force resulting from the pressure is

$$\text{Work done} = \text{force} \times \text{distance}$$

$$= pA \times L$$

The work done per unit mass of the fluid moved is

$$\frac{\text{work done}}{\text{mass } m} = \frac{pAL}{m}$$

where m is the mass of fluid moved.

$$m = AL\rho$$

where ρ is the density of the fluid. The volume moved is AL. Therefore

$$\text{Work done per unit mass} = \frac{p}{\rho}$$

This quantity is sometimes referred to as the *pressure energy* of unit mass.

$$\text{Pressure energy of unit mass} = \frac{p}{\rho}$$

UNIT: J/kg.

The energy possessed per unit mass of fluid will in general be the sum of the pressure energy per unit mass, the *kinetic energy per unit mass* and the *potential energy per unit mass.*

$$\text{Kinetic energy per unit mass} = \frac{\frac{1}{2}mv^2}{m}$$

$$= \tfrac{1}{2}v^2$$

where v is the fluid velocity.

$$\text{Potential energy per unit mass} = \frac{mgh}{m}$$

$$= gh$$

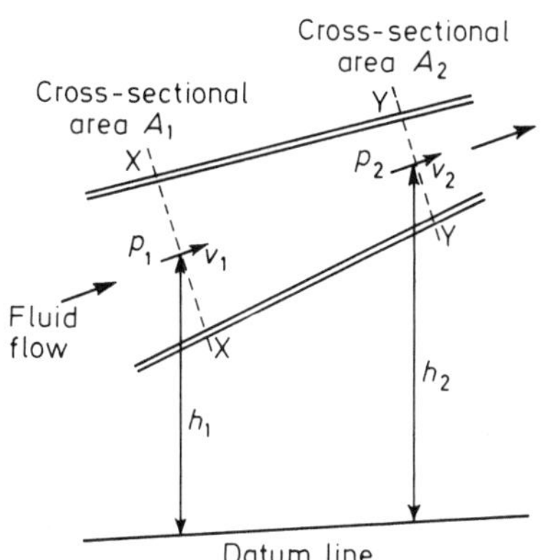

Figure 8.5

where h is the height of the fluid above some datum line. Each of the energy per unit mass terms has the units J/kg.

According to the principle of the conservation of energy the total energy of a fluid will remain constant, provided we take into account all the various forms of energy involved. Thus if a fluid flows through a pipe then the total energy at one section of the pipe must equal the total energy at another section further down the pipe. Hence for the flow through the pipe shown in Figure 8.5,

Total energy at secton XX = total energy at section YY

At XX the pressure energy per unit mass is p_1/ρ, the kinetic energy per unit mass $\frac{1}{2}{v_1}^2$ and the potential energy per unit mass gh_1. At YY the pressure energy per unit mass is p_2/ρ, the kinetic energy per unit mass $\frac{1}{2}{v_2}^2$ and the potential energy per unit mass gh_2. Hence

$$\frac{p_1}{\rho} + \frac{{v_1}^2}{2} + gh_1 = \frac{p_2}{\rho} + \frac{{v_2}^2}{2} + gh_2$$

The above energy equation when expressed in that form is known as *Bernoulli's equation.*

The above equation assumes that no energy is lost, or gained, by the fluid in its movement between sections XX and YY. With steady flow the loss of energy between XX and YY may not be too great, with turbulent flow the loss can be quite significant. The result of any energy loss is that there needs to be added to the energy per unit mass at section YY an energy term representing the energy lost between the two sections. This energy 'lost' will probably have appeared as a rise in temperature of the fluid and the pipe walls.

Bernoulli's equation is sometimes written in a different form. Dividing throughout by g gives

$$\frac{p_1}{\rho g} + \frac{{v_1}^2}{2g} + h_1 = \frac{p_2}{\rho g} + \frac{{v_2}^2}{2g} + h_2$$

The units of each of the above terms are metres.

$$\frac{p}{\rho g} : \frac{\text{N/m}^2}{\text{kg/m}^3 \times \text{m/s}^2} = \frac{\text{N} \times \text{s}^2 \times \text{m}^3}{\text{kg} \times \text{m} \times \text{m}^2}$$

But

$$\text{N} = \text{kg} \times \text{m/s}^2$$

Hence the units are

$$\frac{\text{kg} \times \text{m} \times \text{s}^2 \times \text{m}^3}{\text{kg} \times \text{m} \times \text{s}^2 \times \text{m}^2} = \text{m}$$

$$\frac{v^2}{g} : \frac{(\text{m/s})^2}{\text{m/s}^2} = \frac{\text{m}^2 \times \text{s}^2}{\text{m} \times \text{s}^2} = \text{m}$$

Because each of the terms have units of metres we refer to each as the equivalent head. This is the height at which a static column of liquid would be with the same energy.

(pressure head + kinetic head + potential head) all for the section XX = (pressure head + kinetic head + potential head) all for the section YY.

Example 3 A horizontal pipe tapers uniformly from a diameter of 140 mm to a diameter of 80 mm and carries oil of density 850 kg/m^3. The pressure at the wider diameter section is measured as 80 kPa and that at the smaller diameter section as 50 kPa. Calculate the velocities at both sections and the volumetric rate of flow.

Applying Bernoulli's equation gives

$$\frac{80 \times 10^3}{850} + \frac{v_1^2}{2} + 9.81 \times h = \frac{50 \times 10^3}{850} + \frac{v_2^2}{2} + 9.81 \times h$$

As the height is the same for both sections of the tube that term cancels out from both sides of the equation. Applying the equation of continuity gives

$$A_1 v_1 = A_2 v_2$$

$$\frac{\pi}{4} \times 0.140^2\, v_1 = \frac{\pi}{4} \times 0.080^2 \times v_2$$

$$v_1 = \frac{0.080^2}{0.140^2} \times v_2$$

$v_1 = 0.0327\, v_2$ to three significant figures.

Substituting this in the earlier equation gives

$$\frac{80 \times 10^3}{850} + \frac{(0.0327\, v_2)^2}{2} = \frac{50 \times 10^3}{850} + \frac{v_2^2}{2}$$

Hence

$$35.3 = 0.0499\, v_2^2$$

$$v_2 = 26.6 \text{ m/s}$$

Thus as $v_1 = 0.0327\, v_2$ then $v_1 = 0.870$ m/s.

The volumetric rate of flow is the product Av at any section. Hence

$$\dot{V} = \frac{\pi}{4} \times 0.140^2 \times 0.870$$

$$\dot{V} = 0.0134 \text{ m}^3\text{/s}$$

The above problem indicates a feature common to horizontal tubes: where there is a constriction the velocity increases and the pressure drops.

Example 4 An open water tank discharges water to the atmosphere through a pipe at its base. The height of the water in the tank above the discharge pipe is 5 m. With what velocity will the water emerge through the discharge pipe?

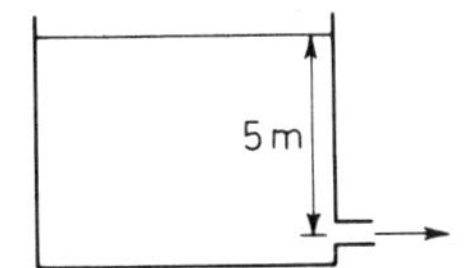

Figure 8.6

Figure 8.6 shows the general arrangement. Applying Bernoulli's equation gives

$$\frac{p}{\rho} + 0 + gh_1 = \frac{p}{\rho} + \frac{v^2}{2} + gh_2$$

The pressure at the open surface in the tank and at the outlet are both the same as they are open to the atmosphere. The velocity at the upper water surface in the tank is assumed to be zero. Hence

$$\frac{v^2}{2} = (h_1 - h_2)g$$

$$v^2 = 2 \times 5 \times 9.8$$

$$v = 9.9 \text{ m/s to two significant figures}$$

Example 5 What will be the volumetric rate of flow from the outlet pipe in the above question if the discharge pipe has a diameter of 100 mm?

$$\dot{V} = Av$$

$$= \frac{\pi}{4} 0.1^2 \times 9.9 \qquad \text{m}^2 \times \text{m/s}$$

$$= 0.078 \text{ m}^3/\text{s}$$

FLOW THROUGH SMALL ORIFICES

When a fluid flows through a small orifice, e.g. a sharp edged hole in the side of a tank, the fluid stream contracts after leaving the orifice and has a cross-sectional area less than that of the orifice. Figure 8.7 shows what happens. The fluid stream contracts to a minimum area at a distance from the orifice of roughly half the orifice diameter. This minimum area is called the *vena contracta*. The ratio of the area of the fluid stream at the vena contracta to the area of the orifice is called the *coefficient of contraction* C_c.

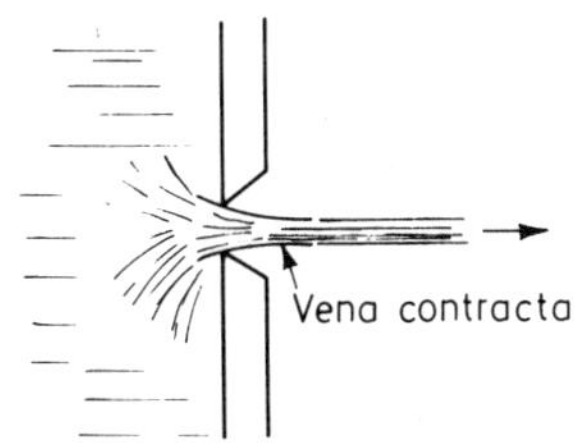

Figure 8.7

$$\text{Coefficient of contraction} = \frac{\text{area of stream at vena contracta}}{\text{area of orifice}}$$

For a small sharp edged orifice C_c has a value of generally about 0.63 to 0.65.

Thus for an orifice with a diameter of 10 mm the area will be $\pi \times 10^2/4$ mm^2 and thus if we take C_c to be 0.65 the area of the vena contracts will be $0.65 \times \pi \times 10^2/4$ mm^2 and thus a diameter d of the fluid stream given by

$$\frac{\pi d^2}{4} = 0.65 \times \frac{\pi}{4} \times 10^2 \qquad \text{mm}^2$$

$$d = 8.1 \text{ mm to two significant figures}$$

For an orifice in the side of a tank of liquid, as in Figure 8.8, we can apply Bernoulli's equation (as in the Example 4 in the previous section) and obtain

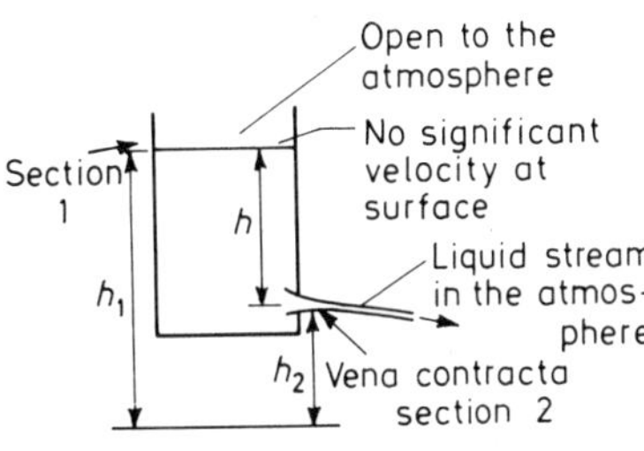

Figure 8.8

$$\frac{p_1}{\rho} \times \frac{v_1^2}{2} + gh_1 = \frac{p_2}{\rho} + \frac{v_2^2}{2} + gh_2$$

The section for which the subscript 1 is used is the open surface of the liquid in the tank, the section for which 2 is used is at the vena contracta. The pressure at the open surface in the tank is the same as at the vena contracta, i.e. the atmospheric pressure. The velocity of the liquid at the surface in the tank is assumed to be zero. The difference in height between the two heights in the equation, i.e. $h = h_1 - h_2$. Hence

$$g(h_1 - h_2) = \frac{v_2^2}{2}$$

or

$$h = \frac{v_2^2}{2g}$$

This gives the velocity of the liquid at the vena contracta if it is assumed that no energy losses are occurring.

Due to frictional effects the velocity at the vena contracta is less than the theoretical velocity calculated from the above equation. The ratio of the actual to the theoretical velocity is called the *coefficient of velocity*, C_v.

$$\text{Coefficient of velocity} = \frac{\text{actual velocity at the vena contracta}}{\text{theoretical velocity at vena contracta}}$$

C_v generally has a value very close to 1, about 0.95.

The theoretical velocity at the vena contracts is given by

$$h = \frac{v^2}{2g}$$

or

$$v = \sqrt{2gh}$$

Therefore the actual velocity will be given by

$$\text{Actual velocity} = C_v \times \sqrt{(2gh)}$$

The volumetric rate of flow $\dot{V}$ for a vena contracta of cross-sectional area A_v will be

$$\dot{V} = A_v v$$

where v is the actual velocity at the vena contracta. Hence using the expression for the theoretical velocity

$$\dot{V} = C_v \times A_v \times \sqrt{(2gh)}$$

But the area of the flow at the vena contracta is related to the area of the orifice A by the coefficient of contraction C_c.

$$C_c = \frac{A_v}{A}$$

Hence

$$\dot{V} = C_v \times C_c \times A \times \sqrt{(2gh)}$$

The product of the two coefficients is generally replaced by another coefficient called the coefficient of discharge C_d.

$$C_d = C_v \times C_c$$

Thus

$$\dot{V} = C_d \times A \times \sqrt{(2gh)}$$

C_d generally has a value of about 0.62. The value depends slightly on the head of liquid in the tank and the shape and condition of the orifice.

Example 6 Water is found to emerge from a 50 mm diameter sharp-edged orifice at the rate of 0.7 m^3/min when the head of water in the tank above the orifice is 5 m. Calculate the discharge coefficient and the rate of discharge when the pressure head is 4 m.

$$\dot{V} = C_d \times A \times \sqrt{(2gh)}$$

Hence

$$C_d = \frac{\dot{V}}{A \times \sqrt{(2gh)}}$$

$$= \frac{0.7/60}{\frac{1}{4}\pi \times 0.05^2 \times \sqrt{(2 \times 9.8 \times 5)}} \qquad \frac{m^3/s}{m^2 \times \sqrt{(m/s^2 \times m)}}$$

$$= 0.60 \text{ to two significant figures}$$

The rate of discharge with a pressure head of 4 m will be given by

$$\dot{V} = 0.60 \times \tfrac{1}{4}\pi \times 0.05^2 \times \sqrt{(2 \times 9.8 \times 4)} \quad m^2 \times \sqrt{(m/s^2 \times m)}$$

The volumetric rate of flow for this head will be $\sqrt{(4/5)}$ times the previous rate.

$$\dot{V} = 0.63 \ m^3/s$$

IMPACT OF JETS ON PLATES

When an object hits a surface, its velocity changes. A velocity change means a momentum change, momentum being the product of the mass of the object and its velocity. For a momentum change to occur there needs to be a force acting on the object, and the rate of change of the momentum equals the applied force (see Chapter 5). For example, when a jet of fluid strikes a plate, there is a change of momentum. There is thus a force applied to the jet to cause it to change its momentum. The jet must however exert a force on the plate, a consequence of Newton's third law of motion.

Consider a jet of fluid, moving with a velocity v, colliding with a stationary plate, the plate being at right angles to the direction of movement of the jet (Figure 8.9). We will assume that the jet hits the plate and loses all its velocity, i.e. it does not bounce back off the plate. Because the fluid is travelling with a velocity v it will cover a distance of vt in time t. Thus all the fluid within the fluid stream within the distance vt of the plate will hit the plate in a time t. This is a volume of fluid of $vt \times A$, where A is the cross-sectional area of the fluid stream. If the density of the fluid is ρ then the mass of fluid hitting the plate in time t will be ρvtA.

Mass of fluid hitting plate in time t $= \rho vtA$

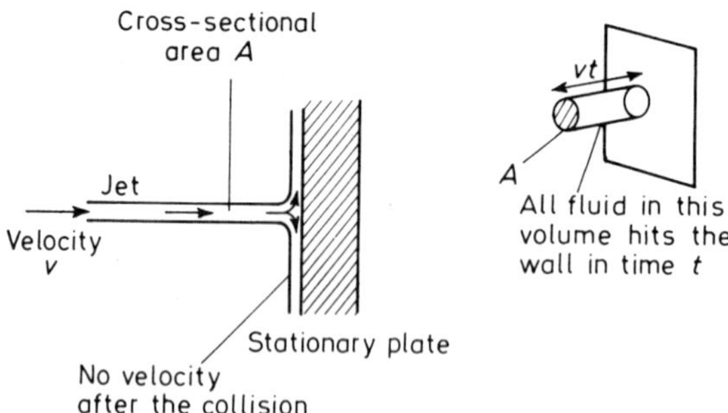

Figure 8.9

As this mass has a velocity v then it must have a momentum given by the product of the mass and the velocity. When the stream hits the plate all this momentum is lost, it having zero momentum after the collision because it is assumed to have zero velocity after the collision. Hence the momentum change occurring in a time t must be

$$\text{Change of momentum in time } t = \rho v t A \times v$$

The rate of change of momentum is therefore

$$\text{Rate of change of momentum} = \rho v^2 A$$

This must be the force acting on the plate.

$$\text{Force } F = \rho v^2 A$$

The situation where a jet of fluid hits a plate is met with in the case of a jet of water hitting the plates of a water wheel (Figure 8.10). The difference between this situation and the argument developed above is

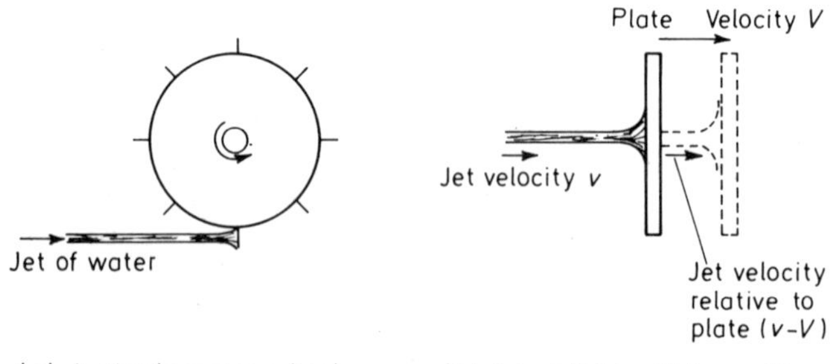

Figure 8.10

that the plates are moving, the whole situation being devised to get the water wheel rotating. If a plate is moving with a velocity V when hit by a water jet moving with a velocity v then the velocity of the jet relative to the plate is $(v - V)$ when the plate is moving in the direction of the jet stream.

$$\text{Mass of water striking plate in time } t = \rho A t(v - V)$$

$$\text{Change in momentum in time } t = \rho A t(v - V) \times (v - V)$$

Hence

$$\text{Force on plate} = \rho A(v - V)^2$$

This force causes the plate to move with a velocity V. In time t the plate will have moved a distance Vt. Hence the work done is

$$\text{Work done in time } t = \text{force on plate} \times \text{distance moved}$$
$$= \rho A(v - V)^2 \times Vt$$
$$\text{Work done per second} = \rho A(v - V)^2 \times V$$

The work done per second is the power. Thus

$$\text{Power} = \rho A(v - V)^2 \times V$$

This energy is obtained from the energy of the jet. The jet, in the form shown in Figure 8.10 has no pressure energy, as it is in the atmosphere, and no change in potential energy. The energy of the jet arises purely from its kinetic energy. The mass of fluid losing kinetic energy per second is ρAv. The kinetic energy $\frac{1}{2}mv^2$ is thus for the jet

$$\text{Energy supplied by jet per second} = \tfrac{1}{2}\rho Av \times v^2$$
$$= \tfrac{1}{2}\rho Av^3$$

The efficiency with which energy is transferred to the water wheel is thus

$$\text{Efficiency} = \frac{\text{work done per second}}{\text{energy supplied by jet per second}}$$

$$\text{Efficiency} = \frac{\rho A(v - V)^2 \times V}{\tfrac{1}{2}\rho Av^3}$$

$$\text{Efficiency} = \frac{2V(v - V)}{v^3}$$

The efficiency is a maximum when $V = v/3$. This is an efficiency of about 30%.

Example 7 A jet of water with a diameter of 300 mm and moving with a velocity of 8 m/s strikes a vane of a water wheel along a line at right angles to the vane. If the vane is (a) stationary, (b) moving with a velocity of 4 m/s in the same direction as the jet, calculate the force exerted by the jet on the vane.

For the stationary vane $F = \rho v^2 A$

The density of water is 1000 kg/m^3, hence

$$F = 1000 \times 8^2 \times \tfrac{1}{4}\pi \times 0.3^2 \qquad \text{kg/m}^3 \times (\text{m/s})^2 \times \text{m}^2$$
$$= 4500 \text{ N to two significant figures}$$

For the moving vane $F = \rho A(v - V)^2$

$$F = 1000 \times \tfrac{1}{4}\pi \times 0.3^2 \times (8 - 4)^2$$
$$= 1100 \text{ N to two significant figures}$$

PROBLEMS

1. The difference in level of the water in a simple manometer is 120 mm. The density of water is 1000 kg/m^3. What is the pressure difference between the two sides of the manometer in pascals?

2. A tank contains water with the water surface being 4 m above the base of the tank. What is the pressure on the base of the tank? The density of water is 1000 kg/m^3.

3. Explain the terms gauge pressure and absolute pressure.

4. Calculate the force on the base of a tank 2.5 m long, 1.5 m wide and filled to a height of 1.2 m with water. The density of water is 1000 kg/m^3.

5. Calculate the volume of water delivered through a pipe of uniform cross-sectional area 800 mm^2 in 5 minutes if the water has a velocity of 10 m/s.

6. Calculate the volumetric rate of flow and the mass flow rate for the water in the previous question. The density of water is 1000 kg/m^3.

7. With what velocity will water need to flow along a pipe of diameter 100 mm if the volumetric rate of flow is to be 0.010 m^3/s?

8. State the equation of continuity for steady flow through a tapered pipe.

9. State Bernoulli's equation and explain the significance of each term in the equation.

10. Water flows down a pipe which tapers from 150 mm diameter to 100 mm diameter at the lower end. The difference in height between the two ends of the pipe is 3 m. The rate of flow is 0.8 m^3/min. Calculate the pressure difference between the upper and lower ends of the pipe.

11. A horizontal pipe carries water, density 1000 kg/m^3. The pipe tapers from 80 mm diameter to 50 mm diameter, the pressure at the 80 mm diameter end being 100 kPa and that at the other end 70 kPa. Calculate the velocity of the water at each end of the pipe and the volume passing through the pipe per second.

12. An open water tank discharges water to the atmosphere through a pipe that falls a distance of 12 m from the water level in the tank. The discharge pipe has a uniform diameter of 50 mm. What will be (a) the velocity with which the water leaves the pipe and (b) the volume discharged per second?

13. A horizontal pipeline 150 mm in diameter has a constriction which reduces the pipe diameter to 90 mm. If water flows along the pipeline with a velocity of 1.2 m/s in the 150 mm diameter part, what will be (a) the velocity at the constriction, (b) the volumetric rate of flow and (c) the mass rate of flow? The water has a density of 1000 kg/m^3.

14. Oil flows through a horizontal pipe which reduces smoothly from a diameter of 75 mm to 50 mm. The gauge pressure at the 75 mm diameter part of the pipe is 70 kPa and that at the 50 mm diameter part 50 kPa. If the density of the oil is 850 kg/m^3, what is the volumetric flow rate and the velocity at that larger diameter section?

15. Oil, of density 850 kg/m^3, discharges through a hole in a tank. The hole has a diameter of 12 mm and the head of oil above the hole is 1.2 m. If the oil discharges at 0.022 kg/s what is the coefficient of discharge for the hole? What would be the rate of discharge in kg/s

if the head increases to 1.5 m? Assume that the coefficient of discharge remains constant.

16. Water emerges from a hole having a coefficient of discharge of 0.64. The hole has a diameter of 25 mm and the head of water above the hole is 3.6 m. Calculate the volumetric rate of flow through the hole.

17. If the head of water in Problem 16 falls to 3.0 m what will be the new volumetric rate of flow through the hole?

18. A jet of water 50 mm in diameter having a velocity of 40 m/s strikes a flat plate fixed at right angles to the jet. Calculate the force acting on the plate. The density of water is 1000 kg/m^3.

19. Calculate the force exerted on a flat plate when a jet of water 60 mm in diameter impinges normally on it with a velocity of 15 m/s. The density of the water is 1000 kg/m^3.

20. Calculate the force exerted on the plate in the previous question if the plate had been moving in the same direction as the jet with a velocity of 7 m/s.

21. What is the energy supplied by a jet moving horizontally through air with a velocity of 20 m/s if the jet stream has a diameter of 40 mm? The density of water is 1000 kg/m^3.

22. Calculate the work done per second when a jet of water of diameter 20 mm strikes a stationary plate with a velocity of 20 m/s. The density of water is 1000 kg/m^3.

23. Calculate the work done by the jet in the previous Problem if the plate, instead of being stationary, is moving with a velocity of 8 m/s in the direction of the jet.

Answers to problems

Answers have not been given to all the problems as lecturers may wish to use some of them for assessment purposes.

CHAPTER 1

4. $T = 33.1$ N, $F = 14.0$ N

5. $T_1 = 7.3$ N, $T_2 = 16.3$ N

8. 44.7 kN at an angle to the 40 kN force of 63° 30′

11. 20 sin 40 and 20 cos 40

13. 3200 N

14. 15 kN in the opposite direction to the force in the cable

16. 2 m from the 2 kg mass. 6 × 9.8 N

18. 2 kN in the direction of the 6 kN force, 18 kN m clockwise.

22. (a) 10 mm above the base along the line bisection the base and joined to the apex. (b) 189 mm along the central axis from the left hand end.

23. Along the central axis 100 mm from one end.

CHAPTER 2

1. (a) 1.6 kN/mm^2, (b) 1.6×10^9 N/m^2, (c) 1.6 GPa

4. 2/300

7. 0.143 mm

11. 79 kN/mm^2, 0.145 kN/mm^2

15. 1 MN

16. (11) 45 N/mm^2

19. 125 N/mm^2

21. Diameter 143 mm, thickness 32 mm

24. 115 kN

26. Steel 68 N/mm^2, copper 39 N/mm^2

28. 85 mm, 4

31. Brass 99 N/mm^2, steel 165 N/mm^2

35. 0.0075 mm

38. 62.5 N/mm^2

41. 140 kN

CHAPTER 3

2. (a) 0.75 kN m, (b) 1.5 kN m, (c) 2.25 kN m. At the centre
3. 24 kN m, at the fixed end
6. 90 kN m, at the fixed end
7. 350 MN/m^2
10. (a) 1.5 MN/m^2, (b) 0.5 MN/m^2
12. 1.97 N m, 79 N/mm^2
15. 52.5 kN
17. 11 kN/m

CHAPTER 4

1. 2.1 mm
3. 60 mm
4. 0.025°
6. 28 mm
10. 4.7 MN/m^2
11. 120 kN/m^2
15. 22 mm

CHAPTER 5

2. (a) 0.2 N, (b) 12 N, (c) 6 N
4. 1.96 N
5. 1.31 N, 3.37 m/s^2
8. 1.02 kg m/s
10. 8 km/h in the direction of the 4000 kg truck
13. 29.4 N
15. (a) 49 N, (b) 42 N, (c) 24.5 N, (d) 0.58
16. 640 km/h in a direction 51.3° east of north

CHAPTER 6

1. $5\pi\ s^{-1}$
2. $2\ s^{-2}$
3. $0.4\pi\ s^{-1}$
7. (a) $0.033\ s^{-2}$ (b) 20 rad
8. $15\ s^{-1}$
11. 10π m/s
14. 42 N m
16. 1.33×10^{-4} kg m^2
19. 29 kN m
21. 1200π W

25. 3400 J, increased by a factor of four
27. 9150 N m, 98 W
32. 13.5 N, increased by a factor of four
36. 8.85 m/s
37. 63°

CHAPTER 7

3. 0.25 Hz
4. 1140 m/s^2, 15 m/s
6. (a) 1.8 Hz, (b) 0.56 s, (c) 10 m/s^2
10. 4.1 Hz
12. 3.5 Hz
16. 4 Hz

CHAPTER 8

1. 1176 Pa
4. 3136 N
5. 2.4 m^3
7. 1.3 m/s
10. 31 kPa
12. 15 m/s, 0.03 m^3/s
15. 0.65
16. 0.0011 m^3/s
18. 3140 N
21. 80.4 kJ
23. 360 J